D1258445

Dictionary of
GLOBAL
CLIMATE
CHANGE

Dictionary of
GLOBAL
CLIMATE
CHANGE

COMPILED BY
W. John Maunder

AS A CONTRIBUTION OF THE
STOCKHOLM ENVIRONMENT INSTITUTE
TO THE SECOND WORLD CLIMATE CONFERENCE

Chapman & Hall
New York

© Stockholm Environment Institute and W. John Maunder 1992

First published in 1992 by UCL Press

UCL Press Limited
University College London
Gower Street
London WC1E 6BT

The name of University College London (UCL) is a registered
trade mark used by UCL Press with the consent of the owner.

First published in North America in 1992 by
Chapman & Hall, Inc.
29 West 35th Street
New York, NY 10001

Printed in Great Britain

Library of Congress Cataloging-in-Publication Data
Maunder, W. J.
   Dictionary of global climate change / compiled by W. J. Maunder as a
contribution of the Stockholm Environment Institute to the Second
World Climate Conference.
     p. cm.
   Includes bibliographical references and index.
   ISBN 0-412-03901-X ; $45.00
   1. Climatic changes. I. Stockholm Environment Institute.
II. World Climate Conference (2nd : 1990 : Geneva, Switzerland)
III. Title.
QC981.8.C5M38   1992
551.6—dc20   92-19059
                                                                    CIP

CONTENTS

FOREWORD

Climate, climate change, climate fluctuations and *climatic trends* are only a few of the terms used today, in not only conferences, scientific symposia and workshops, but also parliaments and in discussions throughout society. To climatologists these terms may be well known; to the vast majority of people, however, they are new, and they require definition and explanation.

The World Meteorological Organization (WMO) inherited an interest and involvement in the studies of climate and climate change from its predecessor, the International Meteorological Organization (IMO), which was established in 1873. By 1929 the IMO had set up a Commission for Climatology to deal with matters related to climate studies. When, in 1950, the World Meteorological Organization assumed the mantle of the IMO, it retained the commission which, among other responsibilities, had already recognized the need for the definition and explanation of terms used in climatology. It must also be said that much of what we now know about climate derives from the scientific and technical programmes co-ordinated by IMO and now, to a much greater extent, by WMO. In 1979, the First World Climate Conference made an assessment of the status of knowledge of climate and climate variability, and recommended the establishment of a World Climate Programme. This recommendation was fully endorsed by the Eighth World Meteorological Congress, and the World Climate Programme was subsequently established by WMO in co-operation with the International Council of Scientific Unions (ICSU) and the United Nations Environment Programme (UNEP). The Second World Climate Conference, convened in October/November 1990, by WMO, and co-sponsored by UNEP, the United Nations Educational, Scientific and Cultural Organization/Intergovernmental Oceanographic Commission (UNESCO/IOC), the Food and Agriculture Organization (FAO) and ICSU, called for some important action programmes, among which were a negotiating mechanism for the development of a Framework Convention on Climate Change, as well as the development of a Global Climate Observing System.

To address the issues relating to terms and terminology used in climatology, in 1990 Dr W. J. Maunder (New Zealand) compiled a provisional edition of "The Climate Change Lexicon", which was made available to participants attending the Second World Climate Conference.

This provided an excellent opportunity to assess the usefulness of such a publication in many allied disciplines. The positive feedback encouraged the revision and expansion of the provisional text prepared by Dr Maunder, who is currently President of the WMO Commission on Climatology. Indeed, Dr Maunder's long experience and broad knowledge in the field of climatology make it befitting for him to author this present new *Dictionary of global climate change*.

Given the importance of this new publication to a wide variety of audiences, I am very happy to convey my thanks and congratulations to the authorities of the Stockholm Environment Institute for their initiative in sponsoring this work. The dictionary will surely contribute to a better understanding of the many complex issues which have arisen and which will continue to arise in the many facets of the climate and climate change arena.

G. O. P. Obasi
SECRETARY-GENERAL
WORLD METEOROLOGICAL ORGANIZATION

PREFACE

The provisional edition of *The climate change lexicon* (renamed in this publication as the *Dictionary of global climate change*) was compiled in 1990 while I was employed by the Stockholm Environment Institute and in the Secretariat of the Second World Climate Conference. The final edition of the *Dictionary of global climate change* was completed during 1991 while I was employed by the Australian Bureau of Meteorology and the Atmospheric Environment Service of Environment Canada, and it reflects comments I received on the provisional edition. It also incorporates many additional items as well as corrections and additions to several of the items which appeared in the provisional edition.

As noted in the Foreword written by Professor G. O. P. Obasi (Secretary-General of the World Meteorological Organization), the original lexicon was written in particular for participants at the Second World Climate Conference. The Co-ordinator of the Conference, Mr H. L. Ferguson, was instrumental in suggesting to me that a lexicon/dictionary of climate change would be very useful, and I would like to highlight the following extracts from the foreword written by Mr Ferguson for the original lexicon:

In 1979 the World Meteorological Organization (WMO) convened a World Climate Conference in Geneva. . . . It reflected a growing interest within the scientific community in the question of climate, climate variations and climate change. In 1986, with public interest in climate change on the rise, WMO decided that it would be appropriate to convene a Second World Climate Conference (SWCC). . . . In the early stages of planning for the SWCC, events in the international arena on global climate change and related problems began to move with unprecedented speed. . . . In particular, the Brundtland Report, issued in 1987, called for new national and international initiatives for sustainable economic development, and the Toronto Conference on the Changing Atmosphere in 1988 called for a "Global Commons" approach and efforts to develop a law of the atmosphere, and proposed quantitative targets for reducing anthropogenic emissions of greenhouse gases. Meanwhile, WMO and UNEP jointly created the Intergovernmental Panel on Climate Change (IPCC). . . .

In the light of this growing "climate of concern" the organizers of the Second World Climate Conference made several important decisions. They decided to hold the Conference in October/November 1990 when the report of the IPCC would be available for review. They determined that the Conference should consist of a scientific/technical component followed by a ministerial meeting, and they assigned an especially high priority to ensuring the participation of the full range of "stake-holders" in the climate change problem. To meet that priority a special effort was made to involve technical experts, planners and policy advisers from a very broad range of disciplines and as many countries of the world as possible. In the event, the Second World Climate Conference attracted over 1,400 participants from 137 countries.

Such a mix presented a sizeable challenge for effective communication. The climate issue, which started from a relatively narrow scientific base, now encompasses a full range of science, technology, legal, environmental, and other socio-economic fields, all accompanied by their jargon, acronyms, and national and international programmes. In an effort to promote better communication among "stake-holders", it was decided that a lexicon of climate and climate change should be prepared and provided to SWCC participants. Through a happy coincidence the Stockholm Environment Institute had under contract Dr John Maunder, President of WMO's Commission for Climatology, and volunteered to make his services available to the SWCC Co-ordination Office.

I hope that the lexicon/dictionary will prove useful not only to those who attended the Second World Climate Conference but also to the media, interpreters and translators, readers of IPCC Reports and other major publications, and the broader community of people interested in climate and climate change.

As will be appreciated, climate and climate change is a rapidly evolving field and I am aware that the final edition is neither complete nor free of error. However, I trust it reflects – at least in part – the position of climate and climate change as it was at the end of 1991.

Various points were raised by the reviewers and others who have independently commented on the provisional edition of, and later additions to, the lexicon/dictionary. Among the comments were those who suggested that the dictionary should be more comprehensive. While that would be desirable, I believe that to make this dictionary much more extensive would require much more expertise than a single author would normally have. Accordingly, it is important to state what the *Dictionary* endeavours to cover and what it does not cover. In particular, while coverage of most climatological and meteorological aspects of climate change has been attempted – in part or in whole – by the various entries, it was not always possible to cover other aspects of climate change in the same comprehensive manner, particularly some of the more biological, ecological, geological, engineering, economic, political and social aspects. In addition, the author is well aware that while most of the important "climate/climate change" activities and programmes of WMO and ICSU are discussed – some in considerable detail – it has not been possible to cover all of the important "climate/climate change" activities and programmes of other international agencies, including those of FAO, UNESCO, IOC, WHO, UNCTAD, UNDRO and UNEP.

Comments from all users of the dictionary are most welcome.

W. JOHN MAUNDER

ACKNOWLEDGEMENTS

It is a pleasure to acknowledge the help I have had from several people and organizations in the preparation of *The climate change lexicon* (renamed in this publication as the *Dictionary of global climate change*). In particular, a special word of appreciation is extended to Dr Gordon Goodman, Executive Director (until 1991) of the Stockholm Environment Institute in Stockholm, Sweden, to Dr Mike Chadwick, Director of the Stockholm Environment Institute at the University of York, UK (until 1991), and Executive Director (from 1991) of the Stockholm Environment Institute in Stockholm, Sweden, to the Swedish Ministry of the Environment, to the Australian Bureau of Meteorology, and to the Canadian Climate Centre of the Atmospheric Environment Service of Environment Canada for financial and/or logistic support. I would also like to thank Mr Howard Ferguson (Co-ordinator of the SWCC) for his initial suggestion that I compile the lexicon/dictionary, and his support during its evolution. I am also appreciative of the help given by many of my colleagues for their support and guidance.

A special word of thanks is given to my wife Melva, who over two years typed onto our personal computer, the many versions of *The climate change lexicon*, including the final edition titled the *Dictionary of global climate change* which you now have in your hands.

W. JOHN MAUNDER
Toronto, Canada
December 1991

SOURCES OF INFORMATION

Because of the complexity of the many items discussed in the *Dictionary of global climate change*, and the fact that most, if not all, terms are modified in some way from their "original" source(s), it is not possible to provide the source(s) of the individual items. However, grateful acknowledgment is made to the following authors, organizations and publishers for the provision of "background" material.

Activities of FAO in the field of climate change (unpublished FAO paper, second draft, December 1991)

Assessing the social implications of climate fluctuations (W. E. Riebsame, under the auspices of UNEP as part of the World Climate Impact Studies Programme, 1989)

The atmosphere and weather of Southern Africa (R. A. Preston-Whyte & P. D. Tyson, Oxford University Press, Capetown, 1988)

Boundary layer climates (T. R. Oke, Methuen, London, 1987)

The changing atmosphere: implications for global security. (Proceedings of the Toronto Conference, 1988)

The climate of Canada (David Phillips, Atmospheric Environment Service, Downsview, Canada, 1990)

The concise Oxford dictionary of Earth sciences (Ailsa Allaby & Michael Allaby (eds), Oxford University Press, Oxford, 1990)

Climate change in the South Pacific (J. W. Zillman, W. K. Downey, M. J. Manton; Scientific Lecture presented at the Tenth Session of the WMO Regional Association V, Singapore, 1989)

Climate change – the New Zealand response (a publication of the New Zealand Ministry of the Environment, Wellington, 1988)

Climate change: a reader's guide to the IPCC Report (a booklet prepared by the Climate Action Network for Greenpeace UK, 1990)

Climate change: meeting the challenge (a report by a Commonwealth Group of Experts, Commonwealth Secretariat, 1989)

Climate change: science, impacts and policy – proceedings of the Second World Climate Conference (J. Jaeger & H. L. Ferguson (eds), Cambridge University Press, Cambridge, 1991)

Climate change: the IPCC scientific assessment (J. T. Houghton, G. J. Jenkins, J. J. Ephraums, Cambridge University Press, Cambridge, 1990)

Climate impact assessment (R. Kates, J. Ausubel, M. Berberian; SCOPE Publication no. 27, John Wiley, Chichester, England, 1985)

Climate variations, drought and desertification (F. K. Hare, WMO Publication no. 653, 1985)

Contemporary climatology (A. Henderson-Sellers & P. J. Robinson, Longman, Harlow, England, 1986)

CRC handbook of chemistry and physics (69th edn) (R. C. Weast (ed.), CRC Press, Boca Raton, Florida, 1988)

Developing policies for responding to climate change (J. Jaeger, a report by the Beijer Institute for the Commonwealth Secretariat, 1988)

Developing policies for responding to climate change (J. Jaeger, for the World Climate Impact Studies Programme of WMO/UNEP, 1988)

A dictionary of Earth sciences (Stella E. Stiegeler (ed.), Macmillan, London, 1976)

A dictionary of the environment (Michael Allaby, Macmillan, London, 1977)

Ecology (C. J. Krebs, Harper & Row, New York, 1972)

Economic and social benefits of meteorological and hydrological services (Proceedings of the WMO Technical Conference, Geneva, 1990)

Energy policy in the greenhouse: from warming fate to warming limit (F. Krause, W. Bach, J. Koomey, Earthscan, London, 1990)

Environment in key words: a multilingual handbook of the environment (I. Paenson, Pergamon Press, Oxford, 1990)

Environmental ecology: the impacts of pollution and other stresses on ecosystem structure and function (Bill Freeman, Academic Press, San Diego 1989)

Environmental glossary (4th edn) (G. W. Frick & T. F. P. Sullivan (eds), Government Institutes Inc., Rockville, Maryland, 1986)

Forty years of progress and achievement – an historical review of WMO (Sir Arthur Davies (ed.), WMO Publication no. 721, 1990)

The full range of responses to anticipated climate change (a report prepared by the Beijer Institute for UNEP, 1989)

Global climate change (a scientific review presented by the World Climate Research Programme, a WMO/ICSU publication, 1990)

The Global Climate Observing System (a proposal prepared by an ad hoc group, convened by the Chairman of the Joint Scientific Committee of the World Climate Research Programme, January 1991)

The global climate system : June 1986–November 1988 (a WMO World Climate Data Programme and UNEP Publication, 1990)

Global ecology: towards a science of the biosphere (M. B. Rambler, L. Margulis, R. Fester (eds), Academic Press, San Diego, 1989)

Global Ocean Observing System: status report on existing ocean elements and related systems (an IOC/WMO publication IOC/INF-833, December 1990)

Glossary: carbon dioxide and climate (a publication prepared by the Carbon Dioxide Information Analysis Center, Oak Ridge National Laboratory, Oak Ridge, Tennessee, 1990)

Glossary of meteorology (R. E. Huschke (ed.), American Meteorological Society, Boston, 1959)

Glossary of terms used in agrometeorology (a publication of World Meteorological Organization, CAgM No. 40, WMO/TD-No. 391, December 1990)

The greenhouse effect, climate change and ecosystems (B. Bolin, B. Doos, J. Jaeger, R. Warrick (eds), Scope Publication no. 29, John Wiley, Chichester, England, 1986)

Greenhouse gas emissions: the energy dimension (a publication of OECD/IEA, Paris, 1991)

GESAMP: two decades of accomplishments (H. L. Windom, International Maritime Organization, London, 1991)

How to slow global warming (D. E. Victor, *Nature* **349**, 451-6, 1991)

The human impact of climate uncertainty. (W. J. Maunder, Routledge, London and New York, 1989; also published in 1990 in Spanish – *El impacto humano sobre el clima* – by Arias Montano Editores, Madrid, Spain)

ICSU year book, 1990.

The impact of climate variations on agriculture, 2 vols (M. Parry, T. Carter, N. Konijn (eds), Kluwer, The Netherlands, 1988)

Integrated Global Ocean Services System: plan and implementation programme 1989-95 (an IOC/WMO report published by the World Meteorological Organization as WMO no. 725, Geneva, 1989)

International glossary of hydrology. (WMO publication no. 385, WMO–UNESCO; 1st edn 1974, 2nd edn in press)

International law of atmospheric protection: a critique of the existing rules (a paper prepared by P. Sands & J. Cameron by the Centre for International Environmental Law, Kings College, London, 1990)

International meteorological vocabulary (a publication of the World Meteorological Organization, Geneva, WMO no. 182. TP 91, 1966)

International negotiations on climate change (a paper prepared by the Center for Global Change, University of Maryland, USA, 1991)

IPCC First Assessment Report: Overview, August 1990

IPCC Report of the Fifth Session, March 1991

IPCC Report of the Sixth Session, October 1991

IPCC Report of Working Group I, June 1990

IPCC Report of Working Group II, June 1990

IPCC Report of Working Group III, June 1990

Living with the lakes: challenges and opportunities (a Progress Report to the International Joint Commission (submitted by the Project Management Team), July 1989)

Man's impact on environment (T. R. Detwyler, McGraw-Hill, New York 1971)

Manual on the terminology of public international law (peace) and international organizations (I. Paenson, published for the Graduate Institute of International Studies, Geneva, by Bruylant, Brussels, 1983)

Meteorological glossary (compiled by D. H. McIntosh, Her Majesty's Stationery Office, London, 1972)

Network Newsletter, vol. 7, no. 1, 1991 (published by UNEP and the Environmental and Societal Impacts Group of the National Center for Atmospheric Research, Boulder, Colorado)

Predicting the Earth's atmosphere: an international challenge (Interim report of the Study Commission of the Eleventh German Bundestag "Preventive Measures to Protect the Earth's Atmosphere", Bonn, 1989)

Proceedings of the World Climate Conference, 1979 (WMO publication no. 537)

Protecting the tropical forests: a high-priority international task (Second report of the Enquete-Commission "Preventive Measures to Protect the Earth's Atmosphere" of the 11th German Bundestag, edited by Deutscher Bundestag, Referat Offentlichkeitsarbeit, Bonn, 1990)

Suggested interpretations of various terms and concepts for purposes of a climate change convention (unpublished internal paper by the Atmospheric Environment Service of Environment Canada, Toronto, 1991)

The uncertainty business: risks and opportunities in weather and climate (W. J. Maunder, Methuen, London, 1986)

Understanding atmospheric change: a survey of the background science and implications of climate change and ozone depletion (Henry Hengeveld, a State of the Environment Report, SOE Report no. 91-2, Environment Canada, Ottawa, 1991)

WMO/UNESCO report on water resources assessment: progress in the implementation of the Mar del Plata Action Plan and a strategy for the 1990s (a WMO/UNESCO publication, 1991)

WMO and UNCED - 1992: protecting the atmosphere, oceans and water resources; sustainable use of natural resources (a publication of the World Meteorological Organization, WMO no. 760, 1991)

WMO Annual Report: 1988, 1989, 1990

WMO Bulletins

WMO Second Long Term Plan: 1988-1997

WMO Third Long Term Plan: 1992-2001

World climatology: an environmental approach (J. G. Lockwood, Edward Arnold, London, 1974)

ABBREVIATIONS & ACRONYMS

ACCAD	WMO Advisory Committee on the World Climate Applications and Services Programme (WCASP) and the World Climate Data and Monitoring Programme (WCDMP)
ACMAD	African Centre of Meteorological Applications for Development
AFOS	Agriculture Forestry and Other Human Activities Subgroup of IPCC WG III (Response Studies).
AGGG	Advisory Group on Greenhouse Gases
AMDASS	Agrometeorological Data System
AOSIS	Association of Small Island States
ASEAN	Association of South East Asian Nations
AWS	automatic weather station
BAPMoN	Background Air Pollution Monitoring Network
CAeM	Commission for Aeronautical Meteorology of WMO
CAgM	Commission for Agricultural Meteorology of WMO
CAS	Commission for Atmospheric Sciences of WMO
CASAFA	Inter-Union Commission on the Application of Science to Agriculture, Forestry and Aquaculture
CBS	Commission for Basic Systems of WMO
CCCO	Committee on Climatic Changes and the Ocean
CCDP	Climate Change Detection Project
CCl	Commission for Climatology of WMO
CEC	Commission for European Communities
CFCs	chlorofluorocarbons
CGIAR	Consultative Group on International Agricultural Research
CHy	Commission for Hydrology of WMO
CIMO	Commission for Instruments and Methods of Observations of WMO
CLICOM	climate-computer system
CLIMAPP	Climate Long-ranged Investigation Mapping and Predictions Project
CMM	Commission for Marine Meteorology of WMO
COADS	Comprehensive Ocean Air Dataset
COBIOTECH	Scientific Committee for Biotechnology
CODATA	Committee on Data for Science and Technology

COSPAR	Committee on Space Research
COWAR	Committee on Water Research
CRU	Climate Research Unit (University of East Anglia, England)
CSERGE	Centre for Social and Economic Research on the Global Environment (University of East Anglia and University College London, England)
CSM	climate system monitoring
DARE	Data Rescue Programme
DBCP	Drifting Buoy Co-operation Panel
EDF	Environmental Defense Fund (USA)
EFTA	European Free Trade Association
EIS	Energy and Industry Subgroup of IPCC WG III (Response Studies)
EMEP	Monitoring and Evaluation of Pollution in Europe
EOS	Earth Observing System
ENSO	El Niño–Southern Oscillation
ERBE	Earth Radiation Budget Experiment
ERS	earth resources satellite
FAGS	Federation of Astronomical and Geophysical Data Sources
FAO	Food and Agriculture Organization
FCCC	Framework Convention on Climate Change
FID	International Federation for Information and Documentation
GACC	General Agreement on Climate Change
GADS	Global Aerosol Data System
GATT	General Agreement on Tariffs and Trade
GAW	Global Atmosphere Watch
GCIP	GEWEX Continental-Scale International Project
GCM	general circulation model
GCOS	Global Climate Observing System
GCTE	Global Change and Terrestrial Ecosystems
GDP	gross domestic product
GDPS	Global Data Processing System
GEDEX	Greenhouse Effect Detection Experiment
GEMS	Global Environment Monitoring System
GESAMP	Joint Group of Experts on the Scientific Aspects of Marine Pollution
GEWEX	Global Energy and Water Cycle Experiment
GFDL	Geophysical Fluid Dynamics Laboratory (USA)
GIEWS	Global Information and Early Warning System on Food and Agriculture
GIS	geographic information system
GISS	Goddard Institute of Space Sciences (USA)

GLOSS	Global Sea-Level Monitoring System
GMCC	Geophysical Monitoring of Climate Change
GNP	gross national product
GOOS	Global Ocean Observing System
GOS	Global Observing System
GO₃OS	Global Ozone Observing System
GRDC	Global Runoff Data Centre
GRID	Global Resources Information Database of GEMS
GCTP	Global Tropospheric Chemistry Programme
GTS	Global Telecommunication System
GTSPP	Global Temperature-Salinity Pilot Project
HABITAT	United Nations Conference on Human Settlements
HDGEC	Human Dimensions of Global Environmental Change
HOMS	Hydrological Operational Multipurpose Subprogramme
HRGC	Human Response to Global Change
HWP	Hydrology and Water Resources Programme of WMO
IAEA	International Atomic Energy Agency
IAU	International Astronomical Union
IBN	International Biosciences Networks
IBP	International Biological Programme
ICID	International Commission on Irrigation and Drainage
ICL	Inter-Union Commission on the Lithosphere
ICOLP	industry co-operative programme for ozone layer protection
ICSTI	International Council for Scientific and Technical Information
ICSU	International Council of Scientific Unions
IDNDR	International Decade for Natural Disaster Reduction
IEA	International Energy Agency
IFAD	International Fund for Agricultural Development
IGAC	International Global Atmospheric Programme
IGAP	International Global Aerosol Programme
IGBP	International Geosphere–Biosphere Programme
IGFA	International Group of Funding Agencies for Global Climate Change
IGOSS	International Global Ocean Services System
IGU	International Geographical Union
IGY	International Geophysical Year
IHD	International Hydrological Decade
IHP	International Hydrological Programme
IIASA	International Institute for Applied Systems Analysis
IJC	International Joint Commission
ILO	International Labour Organization
IMCO	Intergovernmental Maritime Consultative Organization
IMF	International Monetary Fund

IMO	International Meteorological Organization
INC	Intergovernmental Negotiating Committee on a Framework Convention on Climate Change
INC/FCCC	International Negotiating Committee, FCCC
INFOCLIMA	Climate Data Information Referral System
INQUA	International Union for Quaternary Research
INSULA	International Scientific Council for Island Development
IOC	Intergovernmental Oceanographic Commission, and International Ozone Commission
IODE	International Oceanographic Data Exchange
IPCC	Intergovernmental Panel on Climate Change
IQSY	International Years of the Quiet Sun
ISCCP	International Satellite Cloud Climatology Project
ISLSCP	International Satellite Land Surface Climatology Project
ISSS	International Society of Soil Science
ISY	International Space Year
ITCZ	Intertropical Convergence Zone
ITTA	International Tropical Timber Agreement
ITTO	International Tropical Timber Organization
IUB	International Union of Biochemistry
IUFRO	International Union of Forestry Research
IUHPS	International Union of the History and Philosophy of Science
IUMS	International Union of Microbiological Sciences
IUPAC	International Union of Pure and Applied Chemistry
IUPAP	International Union for Pure and Applied Physics
IUPESM	International Union for Physical and Engineering Sciences in Medicine ISY International Space Year
IUBS	International Union of Biological Sciences
IUCN	International Union for the Conservation of Nature and Natural Resources
IUGG	International Union of Geodesy and Geophysics
IUGS	International Union of Geological Sciences
JGOFS	Joint Global Ocean Flux Study
JSC	Joint Scientific Committee for the World Climate Research Programme (WCRP)
JSTC	Joint Scientific and Technical Committee for the Global Climate Observing System (GCOS)
LDC	less developed country
LoA	Law of the Atmosphere
MAB	Man and the Biosphere Programme (Unesco)
MARS	monitoring agro-ecological resources by means of remote sensing and simulation
MDD	METEOSAT Data Dissemination
MECCA	Model Evaluation Consortium for Climate Assessment

MEDI	Marine Environmental Data Referral System
NASA	National Aeronautics and Space Administration (USA)
NCAR	National Center for Atmospheric Research (USA)
NGO	non-governmental organization
NOAA	National Oceanic and Atmospheric Administration (USA)
OAU	Organization of African Unity
OECD	Organization for Economic Co-operation and Development
OHP	Operational Hydrology Programme of WMO
OPEC	Organization of Petroleum Exporting Countries
OZONET	ozone network
PAGES	Past Global Changes Project
PIE	Polar Ice Extent Project
PSA	Pacific Science Association
PSO	Polar Stratospheric Ozone Project
PRs	Permanent Representatives of Members with WMO
SAC	Scientific Advisory Committee for the World Climate Impact and Response Studies Programme (WCIRP)
SAGE	Stratospheric Aerosol and Gas Experiment
SCAR	Scientific Committee on Antarctic Research
SCOPE	Scientific Committee on Problems of the Environment
SCOR	Scientific Committee on Oceanic Research
SCOSTEP	Scientific Committee on Solar–Terrestrial Physics
SEI	Stockholm Environment Institute
SPREP	South Pacific Regional Environmental Programme
SST	sea-surface temperature
START	System for Analysis Research and Training
SWCC	Second World Climate Conference
TCP	Tropical Cyclone Programme
TOGA	Tropical Ocean and Global Atmosphere Programme
TOVS	TIROS operational vertical sounder
TRUCE	Tropical Urban Climate Experiment
TWAS	Third World Academy of Sciences
UNCED	United Nations Conference on Environment and Development
UNCTAD	United Nations Conference on Trade and Development
UNCTC	United Nations Centre on Translational Corporations
UNDRO	United Nations Disaster Relief Co-ordinator
UNECE	United Nations Economic Commission for Europe
UNEP	United Nations Environment Programme
UNESCO	United Nations Educational, Scientific and Cultural Organization
UNIDO	United Nations Industrial Development Organization
UNITAR	United Nations Institute for Training and Research
VCP	Voluntary Co-operation Programme (WMO)

WCAP	World Climate Applications Programme
WCASP	World Climate Applications and Services Programme
WCDMP	World Climate Data and Monitoring Programme
WCDP	World Climate Data Programme
WCIP	World Climate Impact and Response Studies Programme
WCIRP	World Climate Impact Assessment and Response Strategies Programme
WCP	World Climate Programme
WCRP	World Climate Research Programme
WHO	World Health Organization
WIPO	World Intellectual Properties Organization
WMO	World Meteorological Organization
WOCE	World Ocean Circulation Experiment
WWW	World Weather Watch
WWWDM	World Weather Watch Data Management

A

abiotic

Pertaining to the non-living part of an ecosystem or to an environment where life is absent.

ablation

The combined processes (such as melting, wind erosion, sublimation, evaporation and calving) which remove snow or ice from a glacier or from a snow field. Ablation is also used to express the quantity of snow or ice lost by these processes.

absorption

Removal of radiation from an incident solar or terrestrial beam, with conversion to another form of energy: electrical, chemical or heat. The absorption of radiation by the gases of the atmosphere is highly selective in terms of wavelengths and may depend also on pressure and temperature.

absorption band

Range of wavelengths (or frequencies) in the electromagnetic spectrum within which radiant energy is absorbed by a substance.

absorption capacity

In some contexts, the ability of a developing country to accept and utilize the financial and technical support and food aid extended to it.

absorption of radiation

The uptake of radiation energy by a solid body, a liquid or a gas. The energy absorbed is then transferred into another form of energy (usually heat).

ACCAD see **Advisory Committee on Climate Applications and Data of the WMO World Climate Programme**

acclimatization

Acclimatization is the process by which people and animals become adapted to an unfamiliar set of climatic conditions. In the broad, popular sense it implies adjustment to all phases of a new physical and cultural environment, and it is often difficult to distinguish purely climatic phenomena from other factors. In the narrower sense of physiological climatology, acclimatization entails actual changes in the human body brought about by climatic influences. It is associated with a decrease in physiological stress as the body continues to be exposed to the new conditions. Temporary adjustments are made to daily and seasonal weather changes. But, when a person moves to a different climate, a more permanent adaptation gradually takes place. Temperature is the element of greatest significance in acclimatization.

accumulated temperature

The integrated excess or deficiency of temperature measured with reference to a fixed datum over a period of time. If on a given day the temperature is above the datum value for n hours and the mean temperature during that period exceeds the datum line by m degrees, the accumulated temperature for the day above the datum is nm degree-hours or $nm/24$ degree-days. By summing the daily values arrived at in this way, the accumulated temperature above or below the datum value may be evaluated for any period. In practice, daily values of accumulated temperature are not usually derived from hourly values as described above but by a method involving the use of daily maximum and minimum temperatures.

acid rain

Acid rain is the deposition of acids from the atmosphere through rain, snow, fog, or dry particles. The acid in the rain is the result of pollution caused primarily by the discharge of gaseous sulphur oxides and nitrogen oxides into the atmosphere from the burning of coal and oil, during the operation of electricity-generating and smelting industries and from transportation. In the atmosphere these gases combine with water to form acids.

ACMAD see **African Centre of Meteorological Applications for Development**

actinometer

The general name for any instrument used to measure the intensity of radiant energy, particularly that of the Sun. Actinometers may be classified, according to the quantities which they measure, in the following manner:
* pyrheliometers, which measure the intensity of direct solar radiation
* pyranometers, which measure global radiation (the combined intensity of direct solar radiation and diffuse sky radiation)
* pyrgeometers, which measure the effective terrestrial radiation.

actual evaporation

Quantity of water evaporated from an open water surface or the ground.

actual evapotranspiration

The sum of the quantities of water evaporated from the surface and transpired by the vegetation.

adaptation

The adjustment of an organism or population to a new or altered environment through genetic changes brought about by natural selection.

adaptation strategies see management options for responding to climatic change

adiabatic process

A thermodynamic change of the state of a system in which there is no transfer of heat or mass across the boundaries of the system. In an adiabatic process, compression always results in warming, expansion in cooling.

advection

The transport of a property or constituent of the air such as temperature or moisture solely by the motion of the atmosphere. Advection is used to refer to horizontal transport by wind of something carried by the air (e.g. pollutants, heat, fog, etc.).

Advisory Committee on Climate Applications and Data of the WMO World Climate Programme

An advisory committee established by the Executive Council of the World Meteorological Organization (WMO) to oversee the activities of two of the four parts of the World Climate Programme, namely the World Climate Data and Monitoring Programme (WCDMP), and the World Climate Applications and Services Programme (WCASP).

Advisory Group on Greenhouse Gases (AGGG)

The Advisory Group on Greenhouse Gases (AGGG) was established in 1986, jointly by WMO, UNEP, and ICSU, to ensure adequate follow-up to the recommendations of the 1985 (Villach) Conference on the Assessment of the Rôle of Carbon Dioxide and other Greenhouse Gases in Climate Variations and Associated Impacts. In 1988, three special Working Groups were established by the AGGG dealing with (1) Responding to Climate Change: Tools for Policy Development; (2) Options for Reducing Greenhouse Gas Emissions; (3) Targets and Indicators of Climate Change; and (4) Usable Knowledge for Managing Global Climatic Change. Reports from these working groups were prepared by the Stockholm Environment Institute and published in October 1990.

aerobiology

The study of living organisms in the air, ranging in size from viruses, through spores and seeds, to insects and birds.

aerology
The study of the free atmosphere.

aerosol propellant
A liquefied or compressed gas in a container, whose purpose is to expel from the container liquid or solid material different from the aerosol propellant.

aerosols
Aerosols are tiny particles that cause haziness. Aerosols may be either natural or anthropogenic, and most if not all anthropogenic aerosols are usually considered to be pollutants. They are mostly composed of water and pollutants such as sulphuric acid and sea salt. Aerosols in the troposphere are usually removed by precipitation. Aerosols carried into the stratosphere usually remain there much longer. Stratospheric aerosols, mainly sulphate particles resulting from volcanic eruptions, may reduce insolation significantly. About 30% of tropospheric dust particles are the result of human activities. Aerosols are important in the atmosphere as nuclei for the condensation of water droplets and ice crystals, as participants in various chemical reactions, and as absorbers and scatterers of solar radiation, thereby influencing the radiation budget of the Earth–atmosphere system, which in turn influences the climate on the surface of the Earth.

aerosols, global distribution of
A stratospheric reservoir, which is occasionally enhanced by volcanic explosions, exists on a global scale. Aerosols in the upper troposphere may have a significant component derived from the stratospheric aerosol reservoir, particularly following a major volcanic injection. The stratospheric aerosol of the Arctic regions has an important rôle in the formation of polar stratospheric clouds and in the seasonal depletion of ozone in those regions. Under normal, non-volcanic conditions, tropospheric aerosols make up the bulk of columnar abundance. Aerosol loading reaches its maximum in spring and the minimum in winter in both hemispheres.

Based on extensive continental and some marine data, it is generally agreed that the tropospheric aerosol consists of two main components. One component consists of a submicrometre aerosol (often called "fine particles") with a mass mode that varies in geometric mean diameter by mass from 0.1μm to 0.3μm. The sources of these particles are low-temperature chemical reactions in the atmosphere and high-temperature combustion. The second aerosol component, called the coarse mode, has a mass mode diameter between 5μm and 20μm for the continental areas. These particles are produced almost totally by mechanical processes (windblown soil dust, sea salt, road dust, etc.).

aerosols, global sources of

The major sources of natural aerosols are:

* crustal soils (e.g. dust, iron oxides), and the ocean surface (e.g. sea salt);
* products of gas-phase chemical reactions in the atmosphere, with the precursor gases originating from combustion or from biological activity;
* volcanic eruptions, which inject sulphur gases into the stratosphere, with the subsequent formation of sulphuric acid aerosols;
* soot from natural fires;
* atmospheric water clouds.

The major sources of anthropogenic aerosols are:

* burning of fossil fuels and industrial activity, which produce particles directly (i.e. soot, fly-ash, etc.) and also large quantities of oxides of nitrogen and sulphur, which are eventually converted to nitrate and sulphate aerosols;
* slash-and-burn agriculture (soot and oxides of nitrogen);
* mineral aerosols, as a result of poor land-use practices.

afforestation

The conversion of a non-forested ecosystem to a forest by the planting of trees.

AFOS

The Agriculture Forestry and Other Human Activities Subgroup of IPCC WG III (Response Studies).

African Centre of Meteorological Applications for Development (ACMAD)

Following widespread drought in Africa in the 1970s and its significant consequences which affected many African countries, various decisions and resolutions by regional and international organizations led to the establishment of the African Centre of Meteorological Applications for Development (ACMAD).

The long-term objectives of ACMAD is to contribute towards the socio-economic development of African countries, particularly by assisting them in their efforts to attain self-sufficiency in food production, water-resource management, and energy. ACMAD will, in particular, strengthen the capabilities of the national meteorological services and develop their manpower resources in the application and use of meteorological and climatological data, act as a "centre of excellence" in meteorology with respect to scientific research and training, provide a continental watch system, and provide practical applications method-ologies, in order to contribute to improvements in various weather-related activities.

The primary functional areas are: applications development in agrometeorology, climatology and hydrometeorology; meteorological operations on a continental basis; numerical weather analysis and prediction development; and the meteorological data-processing and database facilities.

In June 1989, the second meeting of the Board of Governors was held in Addis Ababa to review the progress made in the preparation of the donors' meeting and in the ratification of the ACMAD constitution as well as the payment of contributions by member states. The Government of Niger has made available to the Centre a building which comprises 15 offices and a conference room, an allocation for the purchase of office equipment and vehicles, and a six-hectare plot of land for its permanent headquarters. The Government has also been very active in encouraging ECA member states and donor countries to support ACMAD. The Centre began operations in 1990.

afterburner

An exhaust gas incinerator used to control emissions of particulate matter.

Agenda 21

Agenda 21 is a "blueprint" for action in all major areas affecting the relationship between the environment and the economy, produced at the United Nations Conference on Environment and Development (UNCED) held in Brazil in June 1992. Agenda 21 focuses on the period up to the year 2000 and extends into the 21st century.

Agenda 21: WMO proposals

* *Free unrestricted exchange of environmental data and information* Countries should promote the process of free, unrestricted exchange of data and information related to the natural environment.
* *Strengthening of national agencies*
  Countries should promote the strengthening of national atmospheric, hydrological, oceanographic, and geophysical agencies to enable those agencies to undertake studies of the natural environment, make adequate systematic observations over areas within national jurisdictions, and to contribute to measurements of the global commons.
* *Development of early warning system*
  The relevant UN bodies, in co-operation with countries and NGOs should promote co-operation in the development of early warning systems concerning changes in the environmental systems, including the atmosphere, ocean, land, and fresh water.
* *Understanding natural environment in its entirety*
  The relevant UN bodies, countries and NGOs should, in organizing systematic observations and research programmes, recognize the complex interrelationships among environmental components including cycles of water, energy and various substances (e.g. carbon).

* *Priority areas for research*
  The relevant UN bodies in co-operation with governments, industry, research institutions, and NGOs should identify priority areas of scientific, technical, and socio-economic research on behaviour and response of components of the natural environment under stress due to human activities.
* *Strengthening International activities on observations and research*
  Relevant UN agency activities in co-ordinating, standardizing and organizing observational programmes, including data exchange and research studies should be strengthened.
* *Partnership of environmental and developmental agencies*
  International agencies responsible for providing authoritative scientific information on the main components of the global environment should be included as equal partners with economic and environmental protection agencies, in co-ordination mechanisms for environmentally sustainable development.

AGGG see **Advisory Group on Greenhouse Gases**

agricultural/climate impacts

Most forms of agriculture are sensitive to climate impacts. This is especially true of rain-fed crop production, which comprises about 80% of global agricultural land-use, and produces 50–60% of all agricultural products. Crop yield – the amount of useful biomass produced by a crop during a growing season – and the impacts of yield changes on farm income, food supply, and the broader economy, are of greatest concern. Changes in climate can affect crop yields directly. However, it is also important to recognize that agricultural production is based on three broad types of resources that include, but go beyond, climate: natural resources (e.g. climate, soil, topography, genetic endowments); capital resources (e.g. fertilizers, animal and machinery power); and human resources (e.g. labour inputs, management practices, market conditions).

Climate can affect capital and human resources as well as the natural environment. For example, wet conditions can hinder cultivation of fields by animals and tractors, and thus delay planting. Particular climatic conditions can also reduce the efficacy of fertilizers, herbicides and other human inputs, thus decreasing crop yields.

agricultural drought

Agricultural drought occurs when rainfall amounts and distribution, soil-water reserves and evaporation losses combine to cause crop or livestock yields to diminish markedly.

agricultural meteorology

The study and application of meteorology and climatology (including weather forecasting) to the specific problems of agriculture. Agriculture is the production of food and fibre in all its forms – crops for human and animal consumption, pasture and range for animal grazing, and crops (including wool and wood) as raw materials for manufactured products. Hence, agricultural meteorology (and/or agricultural climatology) deals with the meteorological and climatological aspects of farming, ranching, and forestry, as well as with substances required for production such as water for irrigation, fertilizer, and agricultural chemicals, and the transportation of the products to markets.

agroforestry
The plantation-like cultivation of trees which are integrated into an agricultural system for the production of wood and other forest products; a system in which the cultivation of trees is combined with field crops or pasture in an ecologically, technically and economically sustainable manner.

Agrometeorological Data System
The Agrometeorological Data System (AMDASS) is the agrometeorological database of FAO, originally constituted as the contribution of the agroclimate unit to the Agroecological Zones Project. The database has been growing constantly through exchanges with other databases and through the systematic collection of climatic bulletins and monographs. AMDASS was recently transferred to and consolidated under CLICOM, the climate computer database system of the WMO World Climate Programme. The stations now number 17,000 world wide, with a focus on developing countries. Three sets of data are given particular attention:
* observed or computed monthly normals for 11 elements;
* long-term monthly rainfall series;
* ten-daily precipitation from selected stations in the Sahel, eastern and southern Africa.

agrometeorology
Agrometeorology or agricultural meteorology is concerned with the interactions between meteorological and hydrological factors and agriculture in the widest sense, including horticulture, animal husbandry and forestry.

air conditioning
Rarely can site selection, orientation, materials and design create the desired indoor climate at all times. In popular usage, the term "air conditioning" has sometimes been restricted to the artificial cooling of the interiors of buildings. In the broader sense, it includes all attempts to modify indoor temperature, humidity, air movement, and composition of the air by artificial means. The demands placed upon air conditioning in the control of these elements depend on the outdoor climate, building design and its related factors, and the kind of indoor climate required. In residential buildings, the functions of different

rooms influence requirements for comfort. Most important, individuals differ widely in their perception of comfort. An air-conditioning system may also cleanse the air of pollutants. Dust, soot, pollens and other solid materials are removed by means of filters, by "washing" the air in a spray chamber, or by precipitating the particles on electrically charged screens.

air mass
A body of air in which the horizontal gradients of temperature and humidity are relatively slight and which is separated from an adjacent body of air by a more or less sharply defined transition zone in which these gradients are relatively large. The horizontal dimensions of air masses are normally hundreds or even thousands of kilometres.

air pollutant
The presence in the atmosphere of any dust, fumes, mist, smoke, other particulate matter, vapour, gas, odorous substances, or a combination thereof, in sufficient quantities and of such characteristics and duration as to be, or likely to be, injurious to health or welfare, animal or plant life, or property, or as to interfere with the enjoyment of life or property.

air pollution episode
A period of abnormally high concentrations of air pollutants, often due to low winds and temperature inversion, that can cause illness and death.

airborne materials
Substances, living or dead, visible or invisible; solid or gaseous, carried from one place to another by air movement.

airborne particulates
Airborne particulates are suspended matter found in the atmosphere. They include windblown dust, emissions from industrial processes, smoke from the burning of wood and coal, and the exhaust of motor vehicles.

albedo
The albedo is the whiteness of, or degree of reflection of incident light from, an object or material. It is defined as the proportion of incoming solar radiation that is reflected. Snow and cloud surfaces have a high albedo, because most of the energy of the visible solar spectrum is reflected. Vegetation and the sea have a low albedo, because they absorb a large fraction of the energy. Cloud is the chief cause of variations in the Earth's albedo.

albedo changes

It has been postulated that, if the area covered by snow and ice increases, more solar radiation is reflected and the Earth cools; as a result, more precipitation falls as snow, the area covered increases further, and the cycle is repeated in an amplifying cascade until glaciation becomes extensive. Conversely, according to this concept, a decrease of snow and ice cover could lead to a warmer Earth. However, the fact that this has not happened in historical times is a result of the strong domination of climate by the annual radiation cycle, and because the largest regional winter snow or ice anomalies fail to survive the following summer; on the other hand, the reduced insolation in winter always allows new ice and snow fields to form. The physical effect of increased snow cover, in terms of its influence on the Earth's total radiation balance, is probably small in comparison with the effect of colour changes (due, for example, to destruction of forests by fires, or to spread of deserts) in low latitudes, where much more radiation is available to be absorbed or reflected.

Aleutian low

A low-pressure area or depression, centred near the Aleutian Islands in the North Pacific, which is a conspicuous feature of the Northern Hemisphere surface mean-pressure chart in winter. The depression has an average central pressure below 1,000 hectopascals in January, and it represents the aggregate of the many depressions which affect this region in winter.

algae

Simple rootless plants that grow in sunlit waters in relative proportion to the amounts of nutrients available. They provide food for fish and small aquatic animals.

altithermal period

A period of high temperatures, particularly from 8,000BP to 4,000BP which was apparently warmer in summers, when compared with the present, and with the precipitation zones shifted polewards. Also called the hypsithermal period.

AMDASS see **Agrometeorological Data System**

anabatic wind

An upslope wind caused by lower air density along the slope than at some distance, horizontally, from it. The wind is associated with strong surface heating of the slope.

analogue

In synoptic meteorology, a past synoptic situation which resembles the current situation over an appreciable area. Analogues are usually selected from the same time of year as the current situation. The sequence of weather which followed an analogue is sometimes used as the basis of both short-range and long-range forecasting.

analogue climate model
A method of predicting a future climate by considering an historical situation which had features similar to those anticipated in the future.

analogue model
The representation of a physical system (prototype) by an analogous physical system, such that the behaviour of the analogue approximately (or exactly) simulates that of the prototype.

anemogram
The record of an anemograph.

anemograph
An anemometer which gives a continuous record of the time variations of the wind.

anemometer
An instrument for measuring wind speed and direction.

aneroid barometer
A barometer whose sensitive element comprises one or more aneroid capsules.

Antarctic ice sheet
The ice budget of the Antarctic ice sheet is expected to continue to be balanced by snow accumulation and iceberg discharge into the Antarctic Ocean. Further, because the horizontal extent of the ice sheet is limited by the size of the Antarctic continent (unless there is a surge of the West Antarctic ice sheet presently firmly anchored off shore by submarine obstructions), the budget of the Antarctic ice cap under warmer climate conditions is likely to be dominated by more abundant snowfall and increased accumulation, yielding a thicker ice cover and a consequent lowering of mean sea-level of between 20cm and 90cm by the year 2100. Glaciologists tend to the view that the melting of the Greenland ice sheet and snow accumulation on the Antarctic ice sheet will counter-balance over the next century. This would leave a net contribution to sea-level rise of about 20cm, coming mainly from the melting of mountain glaciers.

Antarctic ozone hole
In 1985, a striking and alarming phenomenon was first recognized which

has become known as the "hole in the ozone layer". The "hole" occurs over the Antarctic continent in the spring, during which ozone is depleted for about two months. The feature seems to have occurred first in the late 1970s and there has been a definite trend towards a bigger hole since then. The 1987 hole, which persisted longer than in earlier years, had less than half the normal amount of ozone, and (as of August 1990) was the most severe hole so far observed. It must be noted, however, that the rate of decrease of the total amount of ozone, averaged over a year, is considerably less than the decrease observed in the hole. It is now widely accepted by scientists that the hole is caused by chlorine from CFCs, together with the special meteorological conditions associated with Antarctica.

anthropogenic
Manmade or human induced; the term is usually used in the context of emissions produced as the result of human activities.

anticyclone
A mass of air of higher atmospheric pressures than that in adjacent areas. The air circulates in a clockwise direction in the Northern Hemisphere and in an anti-clockwise direction in the Southern Hemisphere. An anticyclone generally brings settled and usually dry weather.

AOSIS see **Association of Small Island States**

aphelion
The point on the orbit of the Earth (or any other body in orbit about the Sun) which is farthest from the Sun; the opposite of perihelion. At present, aphelion occurs about 7 July, when the Earth is about three million miles farther from the Sun than at perihelion. The seasons in which aphelion and perihelion fall undergo a cyclic variation with a period of about 21,000 years. The date of the aphelion passage is advancing slowly (towards dates later in the calendar year) at a rate of about one half-hour each year.

Applications of Meteorology Programme of WMO
The Applications of Meteorology Programme of WMO comprises three areas of application of meteorological services and information: agricultural meteorology, aeronautical meteorology, and marine meteorology, and it promotes the development of infrastructures and services which are required in all three areas for the benefit of countries. Other "application activities" are also carried out under other WMO programmes, e.g. under the WCP and the Hydrology and Water Resources Programme.

aquifer
A porous soil or geological formation in which water may move for long distances and which yields groundwater to springs and wells.

Archival Survey for Climate History Project
The Archival Survey for Climate History Project is to examine the feasibility of using parameteorological data from national archives for filling gaps in the historical climate record. The first phase of the project comprises the gathering of relevant information and data relating to weather and climate phenomena from national archives (such as diaries, letters, church records). WMO, UNESCO, ICSU and the International Council of Archives are collaborating in this project. If the scientific analysis is generally positive, then the project would proceed as part of the WMO Climate Change Detection Project, with possible linkages to the PAGES project of the International Geosphere–Biosphere Project, and the European Paleoclimate Programme of the European Science Foundation.

A meeting of experts was held in February 1990 to discuss the implementation of the project. Initially, the events that will be searched for include 17th-, 18th- and 19th-century droughts, and severe cold spells, using archival information on crop failures, the extent of sea ice, and other parameteorological data. A pilot study has been undertaken in several European countries to determine the feasibility of this approach as a possible climate history indicator. A tremendous amount of data has been uncovered. This will be analyzed in 1991/92 by a core group of experts and a report will then be issued showing project results, problems encountered and recommendations for future work. The data from this project could well form one of the proxy data inputs to the Climate Change Detection Project.

Arctic haze
Since the 1940s, observers in the Arctic have reported the increasing presence of layers of reddish-brown atmospheric haze. The haze, which is observed primarily during winter and spring when the Arctic air is very calm, consists of industrial aerosols, mostly from Europe and northern Asia, that have been transported long distances into the Arctic by prevailing winds. These aerosols include sooty and acidic particles which increase the net absorption and diffusion of spring sunlight in the lower atmosphere. They also increase the surface absorption of sunlight as they settle on snow and ice. The haze may cause spring temperatures in the Arctic to become slightly warmer. If that happens, changes in hemispheric wind patterns could also occur.

Arctic warming
A phrase used to describe the marked warming of the Arctic area that took place between the 1920s and 1950s, during which the ice-free period increased and mean annual temperatures rose by about 4–10°C. It was primarily caused by the more northerly tracks of the Atlantic and Pacific

depressions carrying moist warm air towards the Pole.

aridity
A condition in which evaporation always exceeds precipitation. It is also used to indicate the characteristic of a climate relating to insufficiency or inadequacy of precipitation to maintain vegetation.

ASEAN Workshop on Scientific, Policy and Legal Aspects of Global Climate Change
At the conclusion of the ASEAN Workshop on Scientific, Policy and Legal Aspects of Global Climate Change held in Bangkok on 19–20 September 1990, participants made a number of recommendations, including the following:
* That awareness of global climate issues be promoted, as well as appropriate attitudes towards the conservation of energy and a less wasteful lifestyle;
* That greater consideration be given by ASEAN countries to the potential for significant growth-enhancing investments in energy efficiency;
* That new mechanisms be created for financing investments in energy-efficient technologies;
* That careful planning and proper market incentives and disincentives be used to reach the achievable goal of environmentally sound and sustainable development;
* That where there are threats of serious or irreversible damage to the environment, ASEAN countries base development policies on the precautionary principle, meaning that measures to promote economic development must anticipate and prevent the causes of environmental degradation.

ash
The incombustible solid material released when a substance is burned. The proportion of total ash that emerges into the atmosphere, and also the average size of emerging particles, depend on the velocity of the flue gases. Emerging particle sizes range from 0.2cm downwards, all except the smallest particles being deposited near the source.

Asian–Pacific Seminar on Climate Change, Nagoya City, Japan
The Asian–Pacific Seminar on Climate Change was held in Nagoya, Japan from 23 to 26 January 1991 with participants from 18 countries in the Asian–Pacific Region.
The objectives of the Seminar were to support the effort on global environmental issues, with special attention to the increased greenhouse effect, and enhanced awareness of the implications of climate change in the Asian–Pacific Region. This Seminar was the first opportunity in the

region to review and examine the findings of IPCC as well as those of the Second World Climate Conference (SWCC), in the context of the emerging interests in the region, towards developing regional and international co-operation in responding to the challenge posed by rapid climate change.

The implications of climate change in the Asian–Pacific Region were specifically considered by the Seminar, and the chairman's summary stated that, despite wide diversity in environmental and socio-economic conditions, we are aware that the Asian–Pacific nations have special concern for climate change from the following three viewpoints:

* Regional climate phenomena such as the monsoons, the El Niño Southern Oscillation (ENSO), and tropical cyclones have crucial importance in the region.
* Climate change and associated sea-level rise could have severe adverse impacts on many Asian–Pacific countries, especially on small island nations and low-lying areas, with effects possibly including coastal inundation, decrease of agricultural production, and harm to human health.
* The Asian–Pacific nations, which contain more than half the Earth's population, are crucial in any global effort to limit the rate of climate change, as the Asian–Pacific region currently produces almost a third of global greenhouse gas emissions and is experiencing rapid growth in both population and economic activity. It was noted that per capita emissions and economic resources differ widely among countries in the region.

Association of Small Island States (AOSIS)

Several island nations have voiced concern that their survival depends on the outcome of negotiations to create an international climate change convention. Governments from low-lying islands in the South Pacific, Caribbean, and Indian Oceans representing the Association of Small Island States (AOSIS) were extremely visible and vocal at the Second World Climate Conference (SWCC) in their demands for immediate and significant international cuts in the emission of greenhouse gases. Their alliance is indicative of how climate change negotiations can lead to the development of new coalitions in the international arena.

astronomical theory of climate change

The Earth does not revolve in a circular orbit round the Sun at a constant velocity, but over a period of many years changes take place which affect the amount or distribution of solar radiation received by the Earth. First, the tilt of the Earth's axis of rotation relative to the plane of its orbit varies between 21.8° and 24.4° over a period of 40,000 years. This affects the seasonality or thermal range between summer and winter. Second, the ellipticity or eccentricity of the Earth's orbit varies over a period of about 100,000 years. This results in a greater seasonal range of radiation receipt, notably that when the orbit is at its most eccentric, there will be a 30% difference between aphelion and perihelion, compared with about 7% at present and none when the orbit is circular. Third, the season when the Earth is nearest the Sun (perihelion) changes over a period of about 21,000 years. At present it is nearest in the Northern Hemisphere winter (7 January) and farthest in Northern Hemisphere summer (7 July). It is believed that these cyclic changes may have had considerable importance in "triggering" the start of ice ages. It should be noted that changes in the eccentricity change the total amount of incoming radiation received by the Earth, whereas the other changes mentioned only change the seasonal distribution. (See also **Earth's orbital variations**, and **Milankovitch solar radiation curve**)

atmosphere

The gaseous envelope surrounding a planet. The Earth's atmosphere consists of nitrogen (79.1% by volume), oxygen (20.9% by volume), with about 0.03% carbon dioxide, and traces of argon, krypton, xenon, neon, and helium, plus water vapour, traces of ammonia, organic matter, ozone, various salts and suspended solid particles.

atmosphere/ocean interactions

The dynamics of ocean/atmosphere interactions are understood only partially, and they require an account of the links between the upper and lower layers of the ocean, which are even less understood. However, ocean/atmosphere interaction appears to be an important factor in regard to variations in climate on all timescales, and there are clear indications of its relevance to climate changes on timescales of a few years and longer. Examples of the latter include the Southern Oscillation and the associated El Niño variations, which are associated with significant variations in the seasonal climate in many parts of both hemispheres.

atmospheric dispersion

Atmospheric dispersion is the mechanism which causes the dilution and spread of gaseous or smoke pollution, whereby the concentration is progressively decreased. It is a most important mechanism for the re-distribution of salts, and is relied upon for the removal of the products of combustion. The resulting atmospheric deposition includes washout and rainout, as well as dry deposition.

atmospheric dust

Humans are adding large quantities of fine particles (aerosols) to the atmosphere, from both agricultural and industrial activities. Although most of these aerosols are soon removed by gravity and precipitation, they still affect the radiation balance in the atmosphere. Whether this effect adds to or offsets any warming trend depends on the quantity and nature of the particles as well as the nature of the land or ocean surface below. The regional effects, however, can be significant. For example, high concentrations of sulphur aerosols in the Northern Hemisphere may be causing more low cloud cover (the aerosols increase the condensation rates in clouds), thus reducing sunlight heating. This increased cloud may have temporarily reduced the magnitude of any hemispheric warming that may be occurring due to an enhanced greenhouse effect. However, the net global effects of changes in aerosols are not yet properly understood.

atmospheric stabilization

The term atmospheric stabilization is often used to describe the limiting at certain levels of the concentration of the greenhouse gases. The amount by which human-induced emissions of greenhouse gas must be reduced in order to stabilize at present-day concentrations is substantial. They include carbon dioxide 60–80%, methane 15–20%, nitrous oxide 70–80%, CFC-11 70–75%, CFC-12 75–85%, HCFC-22 40–59%.

atmospheric window

A spectral region in which very little radiation is absorbed by the atmosphere.

atoll

A ring-shaped coral reef appearing as a low, roughly circular (sometimes elliptical or horseshoe-shaped) coral island, or a ring of closely spaced coral islets, encircling or nearly encircling a shallow lagoon in which there is no pre-existing land or islands of non-coral origin. Atolls are surrounded by deep water of the open sea. They are especially common in the western and central Pacific Ocean.

aurora

The luminous, radiant emission from the upper atmosphere over middle and high latitudes, and centred on each of the Earth's magnetic poles. The shimmering, pulsating light of the aurora is often seen on clear winter nights in a variety of shapes and colours.

aurora australis

The aurora of the Southern Hemisphere.

aurora borealis

The aurora of the Northern Hemisphere.

automatic weather station (AWS)

A meteorological station which observes and records measurements of atmospheric pressure, temperature, humidity, precipitation and wind, and transmits them automatically, or on request.

avalanche
The mass of snow and ice falling suddenly down a mountainside and often taking with it soil, rocks and rubble of every description.

average year
A term sometimes used to indicate a year for which the observed hydrological or meteorological quantity approximately equals the long-term average of that quantity.

AWS see **automatic weather station**

Azores high
The semi-permanent subtropical high-pressure area over the North Atlantic Ocean, so named especially when it is located over the eastern part of the Ocean. The same high, when displaced to the western part of the Atlantic, or when it develops a separate cell there, is known as the Bermuda high. On mean charts of sea-level pressure, this high-pressure area is one of the principal centres of meteorological blocking (see **blocking situation**) in mid-northern latitudes.

B

Background Air Pollution Monitoring Network (BAPMoN)
Established in 1968 by the WMO's constituent bodies, BAPMoN was intended to provide continuous information on the Earth's changing atmosphere. Even now, BAPMoN is the only globally operational system for monitoring background atmospheric pollution. BAPMoN monitors the tropospheric atmosphere composition at both baseline and regional levels through a global network of stations. It primarily monitors suspended particulate matter, CFCs, carbon dioxide, methane and atmospheric turbidity. At the end of 1987, BAPMoN operated 196 monitoring sites in 57 countries. At that time, 27 new sites were in preparation.

Base-line monitoring is done at observatory-style stations in remote locations so as to minimize direct regional influences and to provide evenly distributed data at least 60% of the time. Data is collected on a long-term basis. This allows evaluated data to be used to determine trends in the chemical concentrations of various substances in the atmosphere.

BAPMoN see above

baroclinic
A baroclinic atmosphere is one in which surfaces of pressure and density

intersect at some level or levels. The atmosphere is always, to some extent, baroclinic. Strong baroclinicity implies the presence of large horizontal temperature gradients and thus of strong thermal winds.

barograph
A barometer which gives a continuous graphic representation of the atmospheric pressure variations with time.

barogram
The record made by a barograph.

barometer
An instrument for measuring atmospheric pressure.

barometric reduction table
A table used for the reduction of station mercury barometer readings to conditions of standard temperature and standard gravity and, if required, to a standard level (normally mean sea level).

barotropic
A barotropic atmosphere is an hypothetical atmosphere in which surfaces of pressure and density (or specific volume) coincide at all levels. The concept of barotropy gives a useful first approximation in some types of atmospheric situations. The contrasting atmospheric state is the baroclinic.

basal sliding (basal slip)
The movement or speed of movement of a glacier on its bed.

base year
The year used as the base for emissions. Because of variations from year to year it is appropriate to take a three- or five-year period such as 1985–87, or 1985–89 as the "base year".

baseline reference period
The time period over which data inventories of greenhouse gas emissions, which can be used as references against which to measure change in greenhouse gas emissions, are collected. The time period should be long enough to remove possibly large interannual variabilities of inventories, preferably two or three years.

bathymetry
The science of measuring ocean depths to determine the topography of the sea floor.

Beaufort scale
A wind force scale, originally based on the state of the sea, expressed in numbers from 0 to 12.

Bellagio 1987 meeting
A meeting of 25 experts held in Bellagio (Italy) in November 1987 to discuss the strategic policy issues in connection with the greenhouse gas and climate change question. The findings of this meeting of experts were first used as an input to the December 1987 meeting of the Advisory Group on Greenhouse Gases (AGGG). It also had an important input into the "The Changing Atmosphere" conference held in Toronto in June 1988, and the Commonwealth Expert Group on Climate Change and Sea Level. (See also **Villach Technical Workshop 1987**)

benchmark station
A climatological station which is relatively uninfluenced by past or future artificial changes and which provides a continuing series of climatological observations.

benthic organism (benthos)
A form of aquatic plant or animal life found on or near the bottom of a stream, lake or ocean.

Bergen Ministerial Conference (1990)
The Conference stated among other things:
"We welcome the decision of the European communities to establish an Environment Agency and a European Environment Information and Observation Network, charged with the collection of objective, reliable and comparable information at the European level to assist in the effective implementation of environmental policies, and to inform the public on the state of the European environment."
"We also welcome the decision that participation in this Agency should be open on mutually acceptable terms to other countries of the ECE region. We invite UNEP, WMO, ECE and the OECD to co-operate actively in the work of the Agency."

Bermuda high
The semi-permanent subtropical high-pressure area of the North Atlantic, so named especially when it is located in the western part of the Ocean. When displaced towards the eastern Atlantic, it is known as the Azores high. On mean charts of sea-level pressure, this high-pressure area is a principal centre of meteorological blocking (see **blocking situation**) in mid-northern latitudes. Warm and humid conditions prevail over the eastern USA, particularly in summer, when the Bermuda high is well developed and it extends westwards.

bilateral co-operation
This term refers to official development assistance in which one country directly helps another. The two countries are designated as donor and recipient countries.

biodegradable
Any substance that decomposes through the action of micro-organisms.

biodiversity
The diversity of species in a region.

Biological Aspects of the Hydrological Cycle Project (BAHC)
An IGBP Core Project which deals with the problem of resolving the rôle of the biosphere and land-surface processes in the advancement of hydrological models and their integration into climate models.

biological productivity
The amount of organic matter, carbon or energy content that grows or accumulates during a given time period. It usually refers to the growth of plant and/or animal matter in a layer of the ocean or a terrestrial ecosystem.

biomass
The total dry weight of the living organisms of a single species or all species in a community or stand, measured at a particular time. Biomass is comprised of plant biomass (phytomass) and animal biomass (zoomass). The mass of dead and cast-off plant parts is often additionally measured and referred to as dead biomass.

biomass burning
Biomass burning in connection with agricultural/forestry production comprises the combustion of organic waste matter in the field, slash-and-burn shifting cultivation, fuelwood use, and land clearing through forest burning in the course of extending settled areas. Biomass burning can contribute significantly to the global budget of several major trace gases in the atmosphere.

In biomass burning, one has to distinguish between the production of carbon dioxide and the production of other trace gases. To the extent that biomass burning is fed by deforestation, this contributes a net flux of carbon to the atmosphere. If deforestation were compensated for by reforestation or afforestation, this carbon flux would be suppressed, but the releases of the other combustion-product trace gases would still remain.

biomass density
The density of biomass per unit area.

biome

A complex biotic community consisting of all of the plants and animals and their respective communities, including all of their successional phases, in a large geographic area.

biosphere

The part of the Earth and the ocean and the atmosphere in which organisms live. The reservoir of the global carbon cycle that includes living organisms (plants and animals) and life-derived organic matter (litter and detritus). The terrestrial biosphere includes the living biota (plants and animals), and the litter and soil organic matter on land, and the marine biosphere includes the biota and detritus in the ocean.

biosphere reserves

A biosphere reserve is an international designation of recognition from UNESCO under the Man and the Biosphere Programme (MAB). The designation signifies that the area is a good example of some of the ways in which conservation objectives can be balanced with development. The term "biosphere" refers to the association of the designated area within the UNESCO–MAB Programme, and "reserve" means that there are some already-protected sites within the biosphere reserve.

The long-range goal is to create a worldwide network of biosphere reserves to include examples of all the Earth's main ecological systems, with their different patterns of human use and adaptations to them. To receive a designation, each biosphere reserve must have a protected core of undisturbed landscape which can provide baseline data for comparison with nearby areas being managed to meet human needs. As of late-1990, the international network of biosphere reserves comprised 293 sites in seventy-four countries, covering a surface area of some 152 million hectares.

biota

The flora and fauna of an area.

blizzard

A severe winter weather condition characterized by very low temperatures, strong winds and poor visibilities.

blocking high

Any high-pressure area (or anticyclone) that remains nearly stationary or moves slowly compared to the west-to-east motion "up stream" from its location. A blocking high effectively "blocks" the movement of migratory cyclones and depressions, causing them to take a more northerly or southerly path.

blocking situation

The mechanism whereby an anticyclone or high-pressure area remains stationary and cyclones or low-pressure areas move around it. Blocking is the cause of some abnormal weather pattern changes such as extensive drought, prolonged light wind or abnormal wind periods, extended periods of unusual humidity, frostiness, etc.

boreal forest

The conifer forest that occurs in the low Arctic regions with a long and cold winter and a short growing season. It gives way to tundra at more northern latitudes.

boundary conditions

A set of conditions to be satisfied by the solution of a differential equation at the boundary (including fluid boundary) of the region in which the solution is sought.

boundary layer

The atmospheric boundary layer is usually considered as the lowest 1km of the atmosphere, where motion is strongly influenced by surface characteristics, predominantly frictional drag. In the atmospheric boundary layer, surface heat, evaporation and pollutants are very important.

box model

A non-dynamic assessment of the quantity of material or energy in various compartments of an ecosystem, and of the fluxes between the compartments.

bright sunshine

Sunshine intense enough to burn a mark on recording paper mounted in the **Campbell–Stokes sunshine recorder**. The daily period of bright sunshine is less than that of visible sunshine because the Sun's rays are not intense enough to burn the paper just after sunrise, near sunset, and under cloudy conditions.

Bruckner cycle

A climatic cycle of about 35 years, which was deduced by Bruckner in 1890 on the basis of non-instrumental data, such as harvests. He distinguished alternating warm and dry spells, and cold and wet spells, but the cycle is not well defined.

Brundtland Commission

As a consequence of a resolution adopted by the 38th Session of the UN General Assembly, The World Commission on Environment and Development was created in December 1983. The UN Secretary-General appointed Mrs Gro Harlem Brundtland, the then Prime Minister of Norway, as the Chairman of the Commission.

The Commission, which became known as the Brundtland Commission, had a mandate during its three years of existence to re-examine the critical issues of environment and development and formulate new and concrete action proposals to deal with them; to assess and propose new forms of international co-operation that could break out of existing patterns and foster needed change; and to raise the level of understanding and commitment everywhere. During its existence, the Commission functioned as an independent body. It was thus able to address any issues, to solicit any advice, and to formulate and present any proposals and recommendations that it considered pertinent and relevant. The Commissioners each served in a personal capacity and not as representatives of their governments.

The Commission's final report, "Our Common Future", was officially issued in London in April 1987. In October of 1987, the report was presented by the Commission's Chairman, Mrs Brundtland, to a special plenary session of the General Assembly of the United Nations, which subsequently adopted a resolution establishing a broad follow-up procedure on it, with particular emphasis on the UN system and the international financial organizations.

Bureau Gravimétrique International see **Federation of Astronomical and Geophysical Data Sources**

business as usual

This is one of three possible scenarios often used in climate change discussions and policy options which reflect the level of effort and ambition of policies specifically undertaken by governments, companies, and individuals to deal with climate change induced by greenhouse gases. The "business-as-usual" scenario usually implies that no policies specifically directed at limiting greenhouse gas emissions are or will be undertaken.

C

C3, C4 and CAM plants

C3 plants are so named because a compound containing 3 atoms of carbon is the first product of carbon dioxide fixation in photosynthesis. In C4 plants a 4-carbon compound is produced. CAM plants use both pathways (at different times) and are distinctive in being able to fix carbon dioxide in the dark.

C3 plants

Plants such as soya bean, wheat and cotton, whose carbon-fixation products have three carbon atoms per molecule. When compared with C4 plants, C3 plants show a greater increase in photosynthesis for a doubling of CO_2 concentration and less decrease in stomatal conductance, which results in an increase in photosynthetic water-use efficiency.

C4 plants

Plants such as maize and sorghum, whose carbon fixation products have four carbon atoms per molecule. When compared with C3 plants, C4 plants show little photosynthetic response to an increase in CO_2 concentrations above 340ppm but show a decrease in stomatal conductance, which results in an increase in photosynthetic water-use efficiency.

CAeM see **Commission for Aeronautical Meteorology of** WMO

CAgM see **Commission for Agricultural Meteorology of** WMO

CAM plants

CAM (crassulacean acid metabolism) plants are plants such as cactus and other succulents that, unlike the C3 and C4 plants, temporarily separate the processes of carbon dioxide uptake and fixation when grown under arid conditions. They take up gaseous carbon dioxide at night when the stomata are open and water loss is minimal. During the day when the stomata are closed, the stored CO_2 is released and chemically processed. When CAM plants are not under water stress, they follow C3 photosynthesis.

Campbell–Stokes sunshine recorder

An instrument used in many countries for measuring bright sunshine. It consists of a glass sphere 10cm in diameter that is mounted in part of a spherical bowl to which a cardboard card is affixed. The card is scorched by the Sun's rays and measurements of the length of the scorch marks gives the duration of bright sunshine for the day.

canopy

The branches and leaves of woody plants that are formed some distance above the ground.

carbon assimilation

The process by which the carbon contained in atmospheric carbon dioxide is incorporated into the biomass of plants.

carbon budget (balance)

The balance of the exchanges (incomes and losses) of carbon between the carbon reservoirs or between one specific loop (e.g. atmosphere–biosphere) of the carbon cycle. An examination of the carbon budget of a pool or reservoir can provide information about whether the pool or reservoir is functioning as a source or sink for carbon dioxide.

carbon cycle

The carbon cycle consists of the movement of carbon through the biosphere, the Earth's surface and interior, the oceans and the atmosphere. Carbon exists in atmospheric gases, in dissolved ions in the hydrosphere, and in solids as a major component of organic matter and sedimentary rocks. Inorganic exchange is mainly between the atmosphere and hydrosphere. The major movement of carbon results from photosynthesis and respiration, with exchanges between the biosphere, atmosphere and hydrosphere. The rates of exchange are quite small, but over geological time the carbon cycle has concentrated large amounts of carbon in the lithosphere, mainly as limestones and fossil fuels.

Specifically, the carbon cycle consists of all parts (reservoirs) and fluxes of carbon; it is usually thought of as a series of the four main reservoirs of carbon interconnected by pathways of flux. The four reservoirs or regions of the Earth in which carbon behaves in a systematic manner, are the atmosphere, the terrestrial biosphere (usually includes freshwater systems), the oceans and the sediments (includes fossil fuels). Each of these global reservoirs may be subdivided into smaller pools ranging in size from individual communities or ecosystems to the total of all living organisms (biota).

carbon density

The amount of carbon per unit area for a given ecosystem or vegetation type, based on climatic conditions, topography, vegetative cover, type and amount, soils, and maturity of the vegetative stands.

carbon dioxide atmospheric concentrations

Atmospheric measurements of carbon dioxide are now continuously made at 22 Background Air Pollution Monitoring Network stations, and flask samples are taken weekly at another 24 locations. The long-term changes in carbon dioxide concentrations at Mauna Loa Observatory, since routine monitoring began at the Observatory in 1957 when the carbon dioxide concentration was 315ppm, has generally followed the release of carbon dioxide to the atmosphere from the burning of fossil fuels.

Inspection of the measured carbon dioxide record shows three scales of variation: the long-term increasing trend, an annual cycle primarily driven by the Northern Hemisphere vegetative growing season, and a smaller signal usually associated with the El Niño/Southern Oscillation.

The atmospheric carbon dioxide concentration, at 353ppmv in 1990, is now about 25% greater than the pre-industrial (1750–1800) value of about 280ppmv, and higher than at any time in at least the past 160,000 years. Carbon dioxide is currently rising at about 1.8ppmv (0.5%) per year due to anthropogenic emissions. Anthropogenic emissions of carbon dioxide are currently estimated to be about 5.7 ± 0.5Gt C due to fossil fuel burning, plus a range from 0.6 to 2.5Gt C due to deforestation. The atmospheric increase during the 1980s corresponds to $48 \pm 8\%$ of the total emissions during the same period, with the remainder being taken up by the oceans and land. Indirect evidence suggests that the land and oceans sequester carbon dioxide in roughly equal proportions, though the mechanisms are not all well understood. The time taken for atmospheric carbon dioxide to adjust to changes in sources or sinks is of the order of 50–200 years, determined mainly by the slow exchange of carbon between surface waters and deeper layers of the ocean. Consequently, carbon dioxide emitted into the atmosphere today will influence the atmospheric concentration of carbon dioxide for centuries. In order to stabilize concentrations at present-day levels, an immediate reduction in global anthropogenic emissions by 60–80% would be necessary.

carbon dioxide equivalence

The relative rôle of various greenhouse gases in the enhancement of the natural greenhouse effect, either in equivalent units of atmospheric concentrations of carbon dioxide or in units of emissions of carbon dioxide. This term is frequently used as the basis for comparison of the relative greenhouse effect of different greenhouse gases, either in the context of global warming potentials (GWPs) or with respect to changes in atmospheric concentrations of various gases.

carbon dioxide fertilization

Enhancement of growth (or net primary production) due to carbon dioxide enrichment that could occur in natural or agricultural systems as a result of an increase in the atmospheric concentration of carbon dioxide.

carbon dioxide fertilizing effect

A term used to denote the increased plant growth due to higher carbon dioxide concentration in the atmosphere.

Carbon Dioxide Information Analysis Center

A centre established at the Oak Ridge National Laboratory, Oak Ridge, Tennessee, USA, to provide information support to international research, policy, and education communities for the evaluation of complex environmental issues associated with elevated atmospheric carbon dioxide, including potential climate change.

carbon dioxide: sources, amounts, and fluxes

Carbon dioxide, essential to living systems, is released by respiration and removed from the atmosphere by photosynthesis in green plants. The atmospheric content has increased by about 25% since the burning of coal and oil began on a large scale. Atmospheric carbon dioxide varies by a small amount with the seasons, and the oceans contain many times the amount of the gas that is in the atmosphere.

The atmospheric concentration of carbon dioxide, measured at stations such as the Mauna Loa Observatory in Hawaii, is now increasing at the rate of 1.8ppm per year. This annual increase corresponds to the addition of 3.3 billion tons (gigatons) to the total mass of carbon contained in the atmosphere in the form of carbon dioxide. This can be compared with the current annual emission of 5.7 gigatons of carbon in carbon dioxide molecules produced by the combustion of fossil fuels, plus an amount of the order of 1 gigaton resulting from the combustion or oxidation of wood and other biomass. Thus, only about 50% of the anthropogenic carbon dioxide emission remains in the atmosphere. These additional human-induced emissions of carbon dioxide are a small portion of the large fluxes of carbon dioxide between the atmosphere, the ocean, and the terrestrial biosphere during a seasonal cycle, and an even smaller fraction of existing reservoirs of carbon in ocean waters and sediments.

Relatively minor adjustments in the world ocean circulation and chemistry, or in the life cycle of terrestrial vegetation, could affect significantly the amount of carbon dioxide in the atmosphere, even when emissions are stabilized. In particular, ocean warming is likely to decrease the absorption of carbon dioxide by sea water. Conversely, positive changes in the primary biomass production of the ocean could increase the oceanic carbon dioxide uptake and ameliorate the greenhouse effect. Current knowledge of oceanic and terrestrial biochemical processes is not yet sufficient to account quantitatively for exchanges between the atmosphere, ocean and land vegetation: the discrepancy may be as large as 1 gigaton (one billion tonnes) of carbon per year.

carbon flux

The rate of exchange of carbon between the various pools (reservoirs).

carbon intensity

The aggregate measure of carbon emitted as carbon dioxide per unit of activity. It depends on both energy intensity and fuel type.

carbon isotope ratio

Ratio of carbon-12 to either of the other less common carbon isotopes, carbon-13 or carbon-14.

carbon monoxide

A colourless, odourless, very toxic gas produced by any process that involves the incomplete combustion of carbon-containing substances. One of the major air pollutants, it is primarily emitted through the exhaust of gasoline-powered vehicles.

carbon resources

The recoverable fossil-fuel reserves (coal, gas, crude oils, oil shale, and tar sands) and biomass that can be used in fuel production and consumption.

carbon sink

A reservoir that absorbs or takes up released carbon from another part of the carbon cycle. For example, if the net exchange between the biosphere and the atmosphere is towards the atmosphere, the biosphere is the source and the atmosphere is the sink.

carbon source

A reservoir that releases carbon to another part of the carbon cycle.

carbon tax

A tax imposed by an international agency or a country or a local political administrative unit on the amount of greenhouse gas emissions emitted in a particular area.

carbon-based resources

The recoverable fossil-fuel resources (coal, gas, crude oils, oil shale and tar sands) and biomass that can be used in fuel production and consumption.

carrying capacity

The maximum abundance of a species that can occur in an area, beyond which degradation of the habitat will occur.

CAS see **Commission for Atmospheric Sciences of** WMO

CASAFA see **Inter-Union Commission on the Application of Science to Agriculture, Forestry and Aquaculture**

CBS see **Commission for Basic Systems of** WMO

CCCO see **Committee on Climatic Changes and the Ocean**

CCDP see **Climate Change Detection Project**

CCl see **Commission for Climatology of** WMO

CEC
The Commission for European Communities.

Centre de Données Stellaires see **Federation of Astronomical and Geophysical Data Analysis Services**

Centre for Social and Economic Research on the Global Environment (CSERGE)
A research institute investigating the social and economic aspects of global environmental change. Core-funded by the UK Economic and Social Research Council (ESRC), CSERGE is located at two centres: University College London and the University of East Anglia. The 10-year work programme covers the economics, law and politics of global climate change – including the setting of greenhouse gas control targets, the use of economic instruments to achieve those targets, and the equitable distribution world wide of the resulting economic burdens – and the economics, law and politics of biological diversity, especially tropical forest conservation, wildlife management regimes. Other activities include the economics of solid waste management and recycling, the theory and measurement of sustainable development, and the development of the theory and practice of sustainable water management.

CFCs see **chlorofluorocarbons**

CGIAR
Consultative Group on International Agricultural Research

changes in meteorological instrumentation
A major difficulty in assembling homogeneous historical climate datasets which can be used in time-series analyses of the climatic record is adjusting data which have been recorded using different kinds of meteorological instruments. These difficulties arise particularly with respect to precipitation, air temperature, sea temperature, and upper-air temperature observations. The WMO-organized Climate Change Detection Project – among other things – is designed to address this issue.

Changing Atmosphere Conference (Toronto, 1988) see **Toronto Conference (1988)**

chapparal
Vegetation dominated by evergreen angiosperm shrubs, and found in an area with a Mediterranean climate with a hot and dry growing season and a mild and wet winter.

chemical waste

The waste generated by chemical, petrochemical, plastic, pharma-ceutical, biochemical or microbiological manufacturing processes.

chlorofluorocarbons (CFCs)

Chlorofluorocarbons are compounds of carbon containing some chlorine and some fluorine. Chlorofluorocarbons (CFCs) do not occur naturally: they are synthetic products used in various industrial processes and also as a propellant gas for sprays. They are non-poisonous and inert at ordinary temperatures, and are easily liquefiable under pressure, which makes them excellent refrigerants, solvents, foam-makers and for use in aerosol sprays. The uses of CFCs can be regulated and their future abundance depends on the implementation of international agreements to limit production. They have long lifetimes in the troposphere, and how they reach the stratosphere and what happens to them when they reach the stratosphere are important considerations which are currently being studied. Until recently, the largest field of application for CFCs was as propellants for aerosols. Today, CFCs are mainly used as blowing agents in plastics production, as cleaning agents and solvents, and as refrigerants.

There are two main types of chlorofluorocarbons: fully halogenated and partially halogenated. Fully halogenated CFCs are composed exclusively of carbon and halogens, and have a high ozone-depleting potential. Partially halogenated CFCs such as H–CFC-22 (CHF_2Cl) are now being considered as substitutes for fully halogenated CFCs. However, various partially halogenated CFCs are also deleterious in terms of the greenhouse effect, and also contribute – although to a lesser degree – to the destruction of the ozone layer.

chlorofluorocarbon concentrations

The current atmospheric concentrations of anthropogenically produced halocarbons (CFC-11), (CFC-12), (CFC-13), and (CCl4) carbon tetrachloride, are about 280pptv, 484pptv, 60pptv and 146pptv, respectively. Over the past few decades their concentrations, except for CCl4, have increased more rapidly (on a percentage basis) than the other greenhouse gases, the current rate being 4% per year. The fully halogenated CFCs and CCl4 are primarily removed by photolysis in the stratosphere, and have atmospheric lifetimes in excess of 50 years. Future emissions will, most likely, be eliminated or significantly lower than today's because of current international negotiations to strengthen regulations on chlorofluorocarbons. However, the atmospheric concentrations of CFCs 11, 12 and 13 will still be significant (30–40% of current concentrations) for at least the next century because of their long atmospheric lifetimes.

CHy see **Commission for Hydrology of WMO**
CIMO see **Commission for Instruments and Methods of Observations**

CIMO see **Commission for Instruments and Methods of Observations of WMO**

circulation index

A measure of the magnitude of one of several aspects of large-scale atmospheric circulation patterns. Indices most frequently measured represent the strength of the zonal (east–west) or meridional (north–south) components of the wind, at the surface or at upper levels, usually averaged spatially and over time.

city (urban) climates

The urban environment often causes changes in the climate which are sometimes beneficial as in the reduction of humidity due to rain running off surfaces which dry quickly, and an increase of night-time temperature due to use of fuel and trapping of air in streets as well as a reduction of wind strength due to shading by small buildings. However, effects are also sometimes adverse, as in the production of pollution and consequent reduction in sunshine, and the lowering of temperature and generation of fogs, or in the increase of wind caused by tall buildings.

clear-air turbulence

In the free atmosphere, turbulence which is not associated with clouds.

clear cutting

A forest-management technique that involves harvesting all the trees in one area at one time.

clear felling

The complete extraction of the entire standing timber volume of a forest area in a single logging operation.

CLICOM

CLICOM is a WMO activity which aims at assisting mostly developing countries in upgrading their own climate data-processing facilities. Specifically CLICOM is a system which includes three major components: hardware, software and training. Assistance in maintenance is also provided. The hardware is basically any IBM-compatible personal computer with the necessary peripheral equipment. A typical CLICOM station is designed so that it can perform all of the functions of a complete traditional climate data centre. It is therefore designed to perform data entry, quality control, storage and retrieval, data inventories, and basic climatological information products. In the past few years, CLICOM stations have been installed in over 80 countries. WMO's goal is to have this system operational in over a hundred member countries, and by 1992 to have established climate data-management systems in all member countries. Training in the use of the system and in applications of climate analyses to various economic activities in agriculture, energy and water management is an integral part of the project.

CLIMAPP   see **Climate Long-ranged Investigation Mapping and Predictions Project**

CLIMAT messages
Coded messages issued monthly by national meteorological services, which contain surface climatological data (CLIMAT reports) for selected stations. They should normally be issued by the 6th day of each month, and provide data for the previous month.

CLIMAT SHIP
Report of monthly means and totals of various climatic elements from an ocean weather ship.

CLIMAT TEMP
Report of monthly aerological means of various climatic elements derived from upper-air observations over a land station.

CLIMAT TEMP SHIP
CLIMAT TEMP reports from an ocean weather station.

climate
Climate is the synthesis of the day-to-day weather conditions in a given area. The actual climate is characterized by long-term statistics (such as mean values, variances, probabilities of extreme values) of the state of the atmosphere in that area, or of the meteorological elements in that area. Synthesis implies much more than simple averaging. Various methods are used to represent climate; for example, average and extreme values, frequencies of values within stated ranges, and the frequencies of weather types with associated values of elements. The main climate elements are precipitation, temperature, humidity, sunshine and wind velocity, and phenomena such as fog, frost, thunder and gales. Cloudiness, evaporation, grass minimum temperatures, and soil temperatures at various depths, and other items – including those of the upper air – may also be included.

The climate may also be described as the statistical description of weather and atmospheric conditions as exhibited by the patterns of such conditions, in a given region, over a specified period of time long enough to be representative (usually a number of decades).

Climate can also be described as the fluctuating aggregate of atmospheric conditions characterized by the states and developments of the weather of a given area.

climate analogues

Using the notion that "the past is key to the future", climatologists have sought clues on how the greenhouse climate will evolve by referring back to past warm periods. Such analogue approaches generally take one of three forms:

* studies of the board patterns revealed by paleoclimatic reconstructions (for example, warmer conditions associated with the altithermal climate regime that marked the warmest part of the present inter-glacial period about 5,000 to 7,000 years ago are commonly used as an analogue to future greenhouse climates);
* examinations of detailed patterns of warm epochs, usually several years to several decades in length, having occurred recently enough to appear in the instrumental record; or
* studies of groups of individual years (not necessarily consecutive) or seasons selected from the instrumental record for a particular characteristic (e.g. years in which the Arctic was warm).

Such analogues may not accurately reflect future climate. A possible problem with using past warm periods as greenhouse analogues, for example, is that they may not have resulted from greenhouse-like forces, and the forces that shape future climate may yield markedly different climate patterns. In addition, most analogues are chosen on the basis of air-temperature patterns, with no consideration given to other factors that might affect future climate, such as the extent of ice cover or ocean surface temperatures.

climate change

Climate change – in the most general sense – encompasses all forms of climatic inconstancy (that is, any differences between the "long-term" statistics of the meteorological elements calculated for different periods but relating to the same area), regardless of their statistical nature or physical cause.

The "definition" of climate change contained in the Proceedings of the (First) World Climate Conference in 1979 is: "Climate change defines the difference between long-term mean values of a climatic parameter or statistic, where the mean is taken over a specified interval of time, usually a number of decades". This definition implies the existence of climatic change, virtually in all cases where the difference referred to in the definition is not zero. If such a definition is used as a basis for a legal convention on climate change, it will be necessary to assign specific values and statistical significance to the differences between the long-term mean values of the main climatic parameters or statistics, and also to define the beginning and the length of intervals of time over which the mean values are taken.

The term climate change is also often used in the more restricted sense to denote a "significant" change (that is, a change which has important economic, environmental, and/or social effects) in the mean values of a meteorological element (such as temperature or the amount of precipitation) during the course of a certain period of time, where the means are generally taken over periods of a decade or more.

climate change and extreme weather

In spite of the uncertainties associated with them, general circulation models (GCMs) provide a reasonably clear indication of the probable direction of change of the world's average surface climate in the decades to come. However, both humans and ecosystems are much more vulnerable to the occurrence of extreme events such as droughts, floods, heat waves and cold spells than to gradual shifts in climate.

Such an analysis for Canada, for example, suggests that hot spells in the summer will become more frequent and more severe, while very cold periods in winter will become less frequent. In Saskatoon, for example, the frequency of July days exceeding 31°C could increase under a doubled-CO_2 climate from the current average of three days a year to eight. Conversely, extremely cold January days below $-35°C$ could decrease from the current average of three days each year to one day every four years.

climate change as an opportunity: agricultural implications

Temperature currently limits the extension of many crops, such as soya bean and maize. Global warming will make available to agriculture areas which are currently too cold to be cultivated, but this will apply mainly to the developed countries and the Northern Hemisphere. There is also a well known "fertilizing" effect of carbon dioxide. Increased carbon dioxide concentrations are frequently used in real greenhouses markedly to improve yields of high-value horticultural crops. It has also been argued that, at least in part, the constant yield increases achieved with the modern varieties is partially due to the rising atmospheric carbon dioxide. However, it is believed that the fertilizing effect will benefit more those plants adapted to temperate zones (known as C3 species) than those growing in the warmer climates (C4 plants).

Even considering certain negative side-effects (e.g. overwintering of pests), there is a reasonably strong possibility that climate change could become the "climate chance" in many temperate countries in both hemispheres, most of which are in developed countries. It is very unlikely, however, that the tropical and developing countries will benefit in the same measure: indeed, climate change could exacerbate the contrast between the rich and the poor.

climate and energy

By affecting solar radiation receipt, temperature regime, windspeed, water supply, and the growth of some biological fuels (e.g. firewood), climate can influence energy demand and supplies in a region or country. Most importantly, climate determines the thermal environment of human habitations, thus affecting demand for space heating or cooling. It can also affect the search for, and recovery of, energy supplies, especially in harsh environments such as the polar oceans.

climate and fisheries

Both ocean and inland fisheries are sensitive in varying degrees to climate fluctuations. Fish populations are difficult to measure, however, and a lack of data has prevented detailed climate impact assessments in this sector. Yet, both theory and the limited case studies that exist illustrate that certain climate elements do influence fishery productivity, mostly in subtle ways, but at times quite dramatically.

The productivity of fisheries is measured in several ways. Actual populations of fish distributed within age ranges can be estimated by detailed fish sampling (a rare endeavour because it is time-consuming and expensive), or, more frequently, the productivity of a fishery can be measured in terms of total catch in numbers or weight. Climate-induced changes in fish spawning during a certain year (the "year-class" cohort) may affect the catch several years later. Thus, a major problem in correlating catch to climate fluctuation is determining at which stage of their life cycle fish were affected by a particular climatic shift. For species with a long recruitment time it is not unreasonable to postulate that a climate effect occurring up to ten years before the catch was responsible for observed changes in abundance.

Fish populations have shown marked fluctuations throughout recorded history (well before harvesting pressure became significant), and some particular cases illustrate unambiguous connections with climate variation.

climate archive

Central storage of national climatological records in their original form, on microfilms or in digital form in a controlled environment. Climate archive contents include quality-control data, data inventories, station directories, and information on codes, observational practices, and instrumentation.

climate as a common concern

Concerned with the use of climate and the climate resource on the basis that climate is part of the "global commons".

climate-benign technology

Technology which does not cause adverse effects to the climate. Examples would be improved wind- and solar-power generators.

climate – the past 1,000 years

About 1,000 years ago the Earth was in a dry warm period during which the Atlantic Ocean and the North Sea were almost free of storms. This was the time of the great Viking voyages. Vineyards flourished in England, indicating high summer temperatures and mild Mays and Septembers, but there were some frosts in the Mediterranean region, and the occasional freezing of rivers such as the Tiber at Rome and the Nile at Cairo. This suggests that a shift in the pattern of large-scale weather systems was bringing cold Siberian air further south than before.

By AD 1200 the benign climate in western Europe had begun to deteriorate and climatic extremes characterized the next two centuries. Great floods and droughts occurred along with both remarkably severe as well as very warm winters. The Viking colonies in Greenland and Iceland began to deteriorate, and vineyards in both England and Europe fell on hard times. From about 1400 to 1550 the climate grew colder again and from about 1550 the 300-year cold spell known as the "Little Ice Age" began. This period was marked by severe cold weather, with the River Thames in London freezing over with greater frequency than before. The year 1816 was among the worst of many bad years. In much of Europe the winter was wet, the spring was cold and the summer was rainy, and in the United States 1816 came to be known as "the year without a summer". From about 1850, the cold temperatures began to moderate and from about 1890/1900 the long relatively steady warming trend (with a few intervening cold periods) of the 20th century began in many areas of the world.

Climate Change Convention

In 1990, the Executive Heads of WMO and UNEP were authorized by their respective executive bodies to convene an open-ended working group of government representatives to prepare for negotiations on a framework convention on climate change, which took place from 24 to 26 September 1990, in Geneva. Twenty recommendations were adopted by consensus and forwarded to the UN General Assembly for its consideration. The General Assembly in its resolution 25/212 of 21 December 1990, established a single negotiating process under the auspices of the General Assembly, supported by UNEP and WMO, and requested the General Secretary of the UN, in consultation with the Executive Director of UNEP and the Secretary-General of WMO, as well as the Executive Heads of other UN Organizations, to establish at Geneva an ad hoc secretariat, consisting mainly of staff of UNEP and WMO, co-ordinated by those two organizations and supplemented by staff from other bodies of the UN system, as appropriate. The first negotiating session took place from 4 to 14 February 1991, in Washington DC.

Climate Change Detection Project (CCDP)

The WMO Commission for Climatology initiated a Climate Change Detection Project (CCDP) at its Tenth Session held in Lisbon in April 1989. The project received the endorsement of the WMO Executive Council in 1989 and was further discussed at the WMO Congress XI in

May 1991. The project is a worldwide effort, primarily through national meteorological services, to provide reliable analyses of climate trends and variability, and an authoritative evaluation of the climate state for decision-making purposes. To achieve this, the project aims to collect more climate data with well documented station information, and to process them using uniform (objective) procedures.

The CCDP consists of two distinct thrusts: the first is the construction of composite global-climate baseline datasets which include events databases and indices; the second is the utilization of these datasets along with model output predictions to produce, on a regular basis, an authoritative WMO assessment of the status of global climate. These assessments and the information on the global datasets will be distributed to WMO Members. The Project is a natural extension of the ongoing WMO Climate System Monitoring Programme and it fulfils a current need by countries for answers to important national and international concerns.

The project will provide a complete set of climate data with well documented station information to climatologists and other scientists studying the climate in order to detect climate change. The data contained in the set are processed using objective quality-control procedures to enhance their uses by researchers for analyzing climate trends and variability. The set includes cryosphere, ocean, and biomass data integrated with meteorological data, with the resultant dataset being both a point dataset and a gridded dataset. This integrated set will be used as a baseline for anomaly identification and to detect both the primary and secondary signals of climate change. The following specific activities are planned:

* a concerted 10-year effort to integrate high-quality controlled data from various scientific disciplines into a climate database which will be accessible through a geographic information system;
* expert visits to countries to co-ordinate the quality control, exchange and access mechanisms for this dataset;
* co-ordination and training with Members for the use of this dataset as the primary method for detecting climate change;
* preparation of technical guides on climate monitoring and climate trend detection.

As a result of decisions by the Eleventh Session of the WMO Congress meeting in Geneva in May 1991, a Working Group on Climate Change Detection was established, and met for its first session in October 1991. The terms of reference of the working group include the following:

* to prepare regular authoritative reports on the interpretation and applicability of databases for the detection of climate change on regional and global scales and to submit these reports annually to the WMO Executive Council through the President of the WMO Commission for Climatology;
* to serve as an advisory body to the WMO Executive Council through the WMO Commission for Climatology on activities related to the

detection of climate change;
* to provide, as appropriate, an input to updating the implementation plan for the Climate Change Detection Project;
* to co-operate as may be required, with other bodies of WMO and other organizations;
* to keep abreast of scientific developments involving the monitoring and detection of climate change.

At its first meeting in October 1991, the Working Group formed informal groups to address the three main issues concerning the CCDP, namely: quantification of uncertainties and the development of strategies, dataset identification, interpretation and acquisition, and intra-linkages. The Working Group made about thirty recommendations, including the following:
* the need for the preparation of authoritative reports for guidance on the analysis, interpretation, and adjusting of data with errors and biases, which to some extent are known through "metadata";
* climate change detection strategy should be broad but should emphasize the ability to attribute any observed climate changes to greenhouse-gas forcing;
* sponsor an expert meeting on statistical strategies for climate change detection;
* establish advisory groups to prepare guidance material on the construction of metadata and record blending;
* maintain and enhance climate observing programmes;
* establish a dataset certification process through peer-reviewed reports;
* establish a system for giving the "rating" of datasets that takes into account the temporal changes in their quality (e.g. the change from conventional observations to automated systems);
* consult with ICSU and its Panel on World Data Centres to hold "approved" datasets at World Data Centres and establish an international master referral system.

climate change (human induced)

A change of climate directly or indirectly induced by human activities and causing significant environmental, economic or social effects. Any convention on climate change needs to address significant human interference with regional and global climates; hence, any climate convention "definition" of "human-induced climate change" will need to distinguish between natural and anthropogenic, between local and global, and between significant and insignificant changes of the climate.

climate change processes

External processes such as solar irradiance variations, variations of the Earth's orbital parameters (eccentricity, precession and inclination), lithospheric motions and volcanic activity are factors in climatic change. Internal aspects of the climate system also produce fluctuations of sufficient magnitude and variability through the feedback processes interrelating the components of the climate system.

climate changes due to the doubling of carbon dioxide

According to the IPCC Scientific Assessment Report, the main equilibrium changes in climate due to a doubling of carbon dioxide deduced from models are as follows. The number of asterisks (*) indicates the degree of confidence determined subjectively from the amount of agreement between models, our understanding of the model results, and our confidence in the representation of the relevant process in the model. Five asterisks indicate virtual certainties, one asterisk indicates low confidence.

Temperature
*****   the lower atmosphere and Earth's surface warm;
*****   the stratosphere cools;
***   near the Earth's surface, the global average warming lies between +1.5°C and +4.5°C, with a "best guess" of 2.5°C;
***   the surface warming at high latitudes is greater than the global average in winter but smaller than in summer (in time-dependent simulations with a deep ocean, there is little warming over the high-latitude southern ocean);
***   the surface warming and its seasonal variation are least in the tropics.

Precipitation
****   the global average increases (as does that of evaporation), the larger the warming, the larger the increase;
***   increases at high latitudes throughout the year;
***   increases globally by 3% to 15% (as does evaporation);
**   increases at mid-latitudes in winter;
**   the zonal mean value increases in the tropics although there are areas of decrease; shifts in the main tropical rain bands differ from model to model, so there is little consistency between models in simulated regional changes;
**   changes little in subtropical areas.

Soil moisture
***   increases in high latitudes in winter;
**   decreases over northern mid-latitude continents in summer.

Snow and sea ice
****   the area of sea ice and seasonal snow-cover diminish.

The Report added that the results from models become less reliable at smaller scales, so predictions for smaller than continental regions should be treated with great caution.

Climate Data Information Referral System (INFOCLIMA)

The World Climate Data Information Referral Service – INFOCLIMA – is a WMO service for the collection and dissemination of the information on the existence and availability of climate data. INFOCLIMA does not however contain or handle actual climate data.

The information on data sources provided by member countries of the WMO and specialized data centres is collected by the WMO Secretariat and processed into a computerized INFOCLIMA information database which consists of descriptions of available datasets, held at data centres and/or published; an inventory of climatological and radiation stations in the world, together with their historical development, and status surveys of national climatological databanks.

Many sets have been compiled for original observations for the purpose of climate research or for studies in climate applications. Descriptions of sets of original observations and measurements are also included. References are given to publications containing tabulations or charts of climate data. The data descriptions cover a wide range of geophysical subjects, in addition to meteorology, all related to the world's climate system. An interim INFOCLIMA *catalogue of climate system datasets* was published in 1985, and updated in 1990.

The information serves in the planning of regional or global climate studies and in decisions concerning modifications of existing networks or the establishment of new networks for special purposes. Statistics on station networks are issued for WMO Regions.

These databanks are in continuous development, and climate data are being transferred from the older manuscript records to less perishable media and into computer readable forms. Automated data management systems are being introduced to deal with the flow of current data and their processing for climate applications. Status surveys are made regionally to obtain the necessary information for international development projects.

climate discontinuity
A climatic change that consists of a rather abrupt and permanent change during the period of record from one average value to another.

climate fluctuations
Climate varies on all time- and space scales. Large areas of the Earth experience wide variability as part of the "normal" climate. This is especially true of the arid and semi-arid zones, where precipitation varies greatly. Climatic extremes may affect any region; for example, severe drought may occur in humid zones, and floods occasionally occur in dry regions. Any fundamental climate change can mean the establishment of a new "normal" climate to which human activities must adjust.

For analytical purposes, climate fluctuations can be defined as changes in the statistical distributions used to describe climate states. The most commonly cited climate statistic is the simple average of some variable (e.g. temperature) over time. One can imagine that such average values might change by exhibiting unidirectional trends or sharp, step-like changes, increasing or decreasing variability, or a combination of central tendency and variability changes. Although different human activities and resource systems may tend to focus on mean values, the climate impact assessor may be more interested in climate variability events. An increase in variance, even without a simultaneous change in the mean would increase climate stresses on most resource-management systems. Research has shown that small changes in mean values may be associated with marked changes in the frequency of such extremes. Climate impact assessors are also interested in changes in the nature of events typically thought of as comprising weather, rather than climate. For example, whether precipitation tends to occur in gentle, long rains, as opposed to short, heavy downpours, may be critical to some resource-management activities. Other activities may be especially sensitive to certain thresholds of climate conditions. The seasonal distribution of these climate elements can shift without a noticeable effect on mean annual values. Thus, greater consideration of changes in aspects of climate other than mean values – at seasonal as well as annual timescales – may provide better clues to climate impacts.

climate impact assessments

Empirical case studies provide the insight and evidence necessary to understand climate and society interaction and to anticipate future impacts. Candidate cases of climate fluctuations might be selected from the climate record of an area or from well known historical climate events (e.g. a particular drought or extended spell of cold weather). An impact assessor might compare the fortunes of a region experiencing climate fluctuation to those of a nearby, similar region where climate has been more constant. Such a "case-control" approach can isolate climate impacts from the multitude of other physical and socio-economic factors (e.g. soil erosion, pests, market variations, war) that might affect variables selected for analysis (for example, crop yields, farm income, nutrition).

Where such case-control analysis is not feasible, an assessor should take a longitudinal approach and compare indicators before, during and after a climate event. Of course, concurrence of fluctuations and impacts is not in itself proof of a causal connection, but when matched with underlying theory, such coincidence becomes strong evidence of climate–society linkages. Arguments for those linkages can be strengthened further if an assessor examines other, non-climatic explanations for given effects. For example, farm production and income might logically be depressed during a dry spell; however, the same effects might result from increased production costs and decreased crop prices.

climate impact assessments: goals

As well as increasing our basic understanding of climate–society interaction, climate impact studies include the following goals:
* identifying areas, populations or activities particularly sensitive to

climate change;
* measuring and evaluating the impacts of particular climate fluctuations to improve mitigation efforts;
* providing guidance for the application of climate data to improve the resilience of resource-management activities;
* improving the long-range planning of resource management programmes by projecting future climate impacts.

climate impact assessments: sensitivity analysis

Climate impact assessment research shows that climate–society linkages are interactive rather than one-way, cause-and-effect relationships. For example, climate fluctuations may lead directly to poor crop yields or low freshwater runoff, but economic development and human response can either moderate or enhance such problems. Thus, persons assessing climate need to look on the climate–society connection as a process in which the type and magnitude of climate fluctuations interacts with particular social sensitivities to produce impacts which may be ameliorated by subsequent adjustment. Pre-existing sensitivity of human systems is one determinant of the nature and extent of climate impacts, and certain natural and social systems will exhibit a special vulnerability to particular changes. Such differential vulnerability is often neglected in climate impact studies.

climate impact studies

Climate impacts are studied for a wide variety of reasons and from many different perspectives. A government official may need to assess the climate vulnerability of a particular region or economic activity in order to formulate development and investment plans, or may be called upon to assess continuing impacts in order to direct relief aid efficiently. Resource managers must consider climate variation in assessing the short- and long-term productivity of croplands, watersheds, grazing lands, fisheries and forests. Researchers might examine the relationships between climate and society to develop strategies for more efficient use of natural resources and to learn more about how society copes with stress. Projected impacts of potential future climate changes, as distinct from the impacts of climate variations, are also crucial to better economic planning.

Several different approaches to studying climate impacts are available. Assessment types include sensitivity analyses conducted to identify climate vulnerabilities in social systems, empirical case studies focused on certain places or defined climate fluctuations, integrated assessments linking climate impacts to broader socio-economic processes, and projections of likely impacts from future climate changes. The latter approach, often based on statistical models that link climate changes to outcomes in such variables as crop yields, water runoff, or energy use, is especially important because of the need to anticipate the effects of climate warming caused by anthropogenic increases in carbon dioxide and the other greenhouse gases.

climate impacts on energy use

The chief climate impact on energy consumption is related to changes in the demand for space heating and cooling of buildings. Space heating is the principal form of energy consumption in many mid-latitude and northern areas, and space cooling, or "air conditioning", is becoming more widely practised in the subtropics and equatorial regions, and is also particularly important in summer in many developed countries in mid-latitudes. Assessments of climate effects on energy use are dependent on high-quality data, and only in the industrialized countries are aggregate data for different energy systems or distribution grids available over a time period sufficient to determine climate–energy-use relationships. The typical approaches to such analyses are the same as for crop-yield studies, and involve applying either statistical regression analysis, which correlates energy use and temperature, or models that simulate the thermal requirements of individual buildings (or aggregates of buildings) given temperature, wind, sunshine, and other variables. Both approaches demand large amounts of data. Such models are used to estimate the possible changes in energy demand associated with short-term weather fluctuations (in order to predict demands on energy utilities), or to project energy uses under future climate scenarios.

climate impacts on fuelwood supply

In many parts of the world fuelwood is an important source of energy, primarily for domestic use. In fact, in large parts of Africa and southeastern Asia, fuelwood from scrub woodlands is the chief source of energy and is being cut faster than it can regenerate.

The assessment of impacts on fuelwood requires a new research emphasis on the multiple aspects of fuelwood resources, their human management, and their climate sensitivities. Work is especially needed regarding:

* direct climate impacts on woody plant growth, especially in semi-arid woodlands;
* the interaction of climate and fuelwood-harvesting behaviour; and
* fuelwood use, requirements and economics.

climate impacts on hydropower production

The link between climate and hydropower is similar to the link between climate and water. However, the ability to produce hydropower may be even more sensitive to climate-induced water-supply changes than are other water uses, because power-generation turbines operate within a smaller range of water-level variability than do most raw-water supply systems.

Drought has the greatest negative impact on hydropower production. Part of the deficit in power production in such cases is usually due to the need to retain water during certain periods (usually winter) in order to assure adequate supply for summer irrigation. Thus, the goals of hydropower production and drought mitigation often conflict during periods of low precipitation. The degree of this conflict, perhaps expressed as the ratio of water bypassing, versus water passing through, the turbines of a given dam, provides an indicator of the sensitivity of hydropower operations to climate fluctuations.

climate impacts research

A fundamental goal of climate impacts research is to develop a better understanding of the interaction of climate and society so that scientists and policy-makers can predict impacts and formulate responses to future climate changes. Empirical studies are necessary to improve such prediction, to make economic development less vulnerable, and to explain past problems of natural resources management. Despite geographic variability in the mixture of data, tools and institutions available to improve climate impact analysis, the effort must involve the very best minds and must both promote and utilize the co-operation of multiple government agencies and other institutions such as universities. Climate is a fundamental element of the biosphere and it affects, to a greater or lesser degree, almost all human activities; any effort to assess and mitigate its negative effects must be multidisciplinary and broadly collaborative.

Climate Institute

A non-profit climate and climate-change institute located in Washington DC.

Climate Long-ranged Investigation Mapping and Predictions Project (CLIMAPP)

An integrated project to study the climatic history of the Quaternary, conducted by a team of scientists engaged in Earth and ocean research. The administrative base for the project is at Columbia University, New York.

climate model development

The evolution of the multi-faced WCRP modelling activity has led to specialization along two parallel approaches:

The first focus, the WCRP Numerical Experimentation Programme, carried out in co-operation with the WMO Commission for Atmospheric Science is focused on refining the formulation of atmospheric general circulation models, taking into account the results of process studies in the field, the intercomparison of model climatologies and identification of systematic errors, and the day-to-day validation of numerical weather-prediction products against observations. An important activity promoted by the programme is the re-analysis of the past ten years of archived meteorological and oceanic observations, using a single advanced data analysis and assimilation system, in order to produce a consistent climatological series of global atmospheric fields, thus eliminating artefacts caused by frequent changes in the analysis procedure.

The second focus of research is the development of global atmospheric–ocean–ice models, as required for the assessment of natural climate variability and the estimation of transient climate change caused by the progressive increase in the concentration of greenhouse gases. This activity involves a range of model intercomparison studies, beginning with a project to compare ten-year integrations of the atmospheric components only, using a standard set of observed sea-surface temperature and sea-ice conditions.

climate models: confidence in predictions (IPCC WG I: Policymakers Summary)

What confidence can we have that climate change due to increasing greenhouse gases will look anything like the model predictions? Weather forecasts can be compared with the actual weather the next day and the skill of forecasters assessed; we cannot do that with climate predictions. However, there are several indicators that give us some confidence in the predictions from climate models.

When the latest atmospheric models are run with the present atmospheric concentrations of greenhouse gases and observed boundary conditions, their simulation of present climate is generally realistic on large scales, capturing the major features such as the wet tropical convergence zones and mid-latitude depression belts, as well as the contrasts between summer and winter circulations. The models also simulate the observed variability; for example, the large day-to-day pressure variations in the middle-latitude depression belts and the maxima in interannual variability responsible for the very different character of one winter from another both being represented. However, on regional scales (2,000km or less), there are significant errors in all models.

Overall confidence is increased by atmospheric models' generally satisfactory portrayal of aspects of variability of the atmosphere, for instance those associated with variations in sea-surface temperature. There has been some success in simulating the general circulation of the ocean, including the patterns (though not always the intensities) of the principal currents, and the distributions of tracers added to the ocean.

Atmospheric models have been coupled with simple models of the ocean to predict the equilibrium response to greenhouse gases, under the assumption that the model errors are the same in a changed climate. The ability of such models to simulate important aspects of the climate of the last ice age generates confidence in their usefulness. Atmospheric models have also been coupled with multi-layer ocean models (to give coupled ocean–atmosphere GCMs) which predict the gradual response to increasing greenhouse gases. Although the models so far are of relatively coarse resolution, the large-scale structures of the ocean and the atmosphere can be simulated with some skill. However, the coupling of ocean and atmosphere models reveals a strong sensitivity to small-scale errors which leads to a drift away from the observed climate. As yet, these errors must be removed by adjustments to the exchange of heat between ocean and atmosphere. There are similarities between results from the coupled models using simple representations of the ocean and those using more sophisticated descriptions, and our understanding of such differences as do occur gives us some confidence in the results.

climate predictions: uncertainties

Uncertainties in climate predictions arise in part from our imperfect knowledge of the "natural" climate, as well as our imperfect understanding of the future rates of greenhouse gas emissions, how these will change atmospheric concentrations, and the response of climate to these changed concentrations. With regard to greenhouse gas emissions, and their influence on the climate in the future, three factors need to be stressed:

First, future climate changes will depend on the rate at which greenhouse gases (and other gases which influence them) are emitted; this in turn will be determined by various complex economic and sociological factors.

Second, because we do not fully understand the sources and sinks of the greenhouse gases, there are uncertainties in our calculations of future concentrations arising from a given emissions scenario. Because natural sources and sinks of greenhouse gases are sensitive to a change in climate, they may substantially modify future concentrations; ice-core records show that methane and carbon dioxide concentrations changed in a similar sense to temperature between ice ages and interglacials. For example, if wetlands become warmer, methane emissions (and hence concentrations) will increase; if they become drier, methane emissions will decrease. There are also other important processes in the oceans which affect greenhouse gas concentrations. For example, it appears that, as the climate warms, these processes will lead to an overall increase, rather than decrease, in natural greenhouse gas abundances.

Third, models are only as good as our understanding of the natural processes which affect climate, and this is far from perfect. The range in the climate predictions reflects an estimate of uncertainties due to model imperfections; the largest of these is cloud feedback (those factors affecting the cloud amount and distribution, and the interaction of clouds with solar and terrestrial radiation), which leads to a factor of two in the uncertainty over the size of the warming. Others arise from the transfer of energy between the atmosphere and ocean, the atmosphere and land surfaces, and between the upper and deep layers of the ocean.

It must also be recognized that our imperfect understanding of climate processes (and our corresponding ability to model them) could make us vulnerable to surprises; for example, the ozone hole over Antarctica was entirely unpredicted. In particular, the ocean circulation, changes in which are thought to have led to periods of comparatively rapid climate change at the end of the last ice age, is not well understood or modelled.

climate predictions: use of models

There have been many attempts to predict the climatic changes that could be induced by increased levels of carbon dioxide in the atmosphere. One of the earlier models (1964), which looked at a single vertical slice through the atmosphere and assessed how changes in the longwave radiation fluxes affected the temperature profile, was able to show unequivocally that any increase in greenhouse gas concentrations would result in warming of the lower atmosphere and cooling of the stratosphere. As knowledge of the climate system has grown and the speed of computers has increased, it has been possible to build progressively more complex and realistic models.

The most sophisticated climate models today are the so-called general circulation models (GCMs), also sometimes more correctly called "global climate models", which evolved as by-products of global numerical weather-prediction (NWP) models. These models treat the atmosphere as a network of "grid" points over a three-dimensional domain, and solve a set of equations for values of the atmospheric winds, pressure, temperature and humidity at these gridpoints. The difference between GCMs and NWP models is that the weather-prediction models are only concerned with weather developments for a few days ahead, and so can neglect many of the feedback processes that operate over longer timescales. On the other hand, GCMs must model ocean, sea-ice and land-surface processes in considerable detail in addition to treating the day-to-day weather variability.

The various parts of the climate system respond to, or "feedback" on, changes in other parts of the system, and it is because of these feedbacks that the whole global system must be treated together, rather than just focusing on a local area such as Australia. For example, reducing snow and ice cover in high latitudes leads to more solar radiation being absorbed than before, and thus further temperature increases and melting of more ice. Warming the ocean surface leads to more evaporation and, since water is itself a greenhouse gas like carbon dioxide, this causes additional warming of the planet. The ice-albedo and water vapour-temperature effects are examples of "positive feedbacks", whereby any initial warming due to CO_2 increase is magnified.

In developing a model of the global climate, it is inevitable that some compromises will be made in the number of physical processes that can be included and the complexity with which they are represented in the model. There is also a constraint on the "resolution" of the model, which sets a lower limit on the scale of disturbances that can be treated explicitly. All those processes that operate on a smaller scale than the gridpoint separation of the model must be parameterized – that is, the interactions with the larger scale must be expressed in terms of the gridpoint values the model works with. The treatment of clouds is one of the most important of these sub-gridscale effects. The best-documented general circulation models include those developed at the Geophysical Fluid Dynamics Laboratory in Princeton (GFDL), the Goddard Institute for Space Physics (GISS) in Washington, the National Center for Atmospheric Research (NCAR) in Boulder, Colorado, the United Kingdom Meteorological Office, in Bracknell, and at the Canadian Climate Centre in Toronto. The GFDL and NCAR models have a horizontal gridpoint spacing of about 4.5° in latitude and 7.5° in longitude, which means that a small landmass such as New Zealand is virtually "lost" in the surrounding oceans. The GISS model has a coarser resolution (about a 10° spacing between gridpoints), but as compensation for this has more complicated sub-models for the ocean, sea ice, soil moisture and clouds. The GISS model also allows for the diurnal cycle of solar radiation.

A typical general circulation model makes over 500 billion calculations in simulating one year of global climate, and in climate change experiments it is necessary to cycle the predictions through many years of model time. The standard procedure is to first make a "control" run of the model for present-day conditions, and then repeat the calculations for some arbitrarily prescribed change such as increased carbon dioxide concentration. These two simulations are continued for some years until the model reaches "equilibrium" – i.e. there are no significant trends in the model variables (such as surface temperature), although weather variability from daily to interannual timescales will still be present. The simulations are extended a further few years while climate statistics are accumulated. The differences between the equilibrium solutions of the present-day climate and the perturbed climate can then be examined.

climate properties

There are three basic properties that characterize the climate: thermal (surface air temperatures, water, land, ice); kinetic (wind and ocean currents, together with associated vertical motions and the motions of air masses, aqueous humidity, cloudiness, and cloud water content, groundwater, lake lands, and water content of snow on land and sea ice); and static (pressure and density of the atmosphere and ocean, composition of the dry air, salinity of the oceans, and the geometric boundaries and physical constants of the system). These properties are interconnected by various physical processes such as precipitation, evaporation, infrared radiation, convection, advection, and turbulence.

climate reference station see Reference Climatological Station

climate scenarios

Climate scenarios are expectations of what the future climate will be like using as a basis various "what if?" considerations. Many climate scenarios are based on the output of climate models. The most elaborate climate model employed at present consists of an atmospheric global circulation model (GCM) coupled to an ocean GCM which describes the structure and dynamics of the ocean. Added to this coupled model are appropriate descriptions, although necessarily somewhat crude, of the other components of the climate system (namely, the land surface and the ice), and the interactions between them. If it is run for a few years with parameters appropriate to the current climate, the statistics of the model's output (which is a description of the model's climate) will, if the model is a good one, bear a close resemblance to the existing climate which is observed. If new parameters applicable to a changing climate – for instance, changing greenhouse gases are introduced into the model, it can be employed for simulating or predicting climate change.

To run a model of the kind described above requires very large computer resources. Since these have become available only recently and in a few places, simplified versions of the model have commonly been employed to explore the various sensitivities of the climate system and to simulate climate change. In particular, various simplifications of the ocean structure and dynamics have been employed.

The simplest way of employing a climate model to determine the response to a given change in forcing arising, for instance, from an increase in greenhouse gases, is first to run the model for a period (typically several years) with the current forcing; then to change that forcing by, for instance, doubling the concentration of carbon dioxide in the appropriate part of the model description, and run the model again for a period. Comparing the two model climates will then provide an indication of the different climate that is likely to be expected under the new conditions. Such a scenario will be based on an equilibrium response: i.e. it is the response expected from that change when the whole climate system has reached a steady state.

climate sensitivity

The magnitude of a climatic response to a perturbing influence. In mathematical modelling of the climate, it is the difference between simulations as a function of a change in a given parameter.

climate severity index

A number scale from 0 to 100 that describes many of the unfavourable (uncomfortable, depressing, confining and hazardous) aspects of the climate, such as their intensity, duration and frequency of occurrence.

climate signal

Occurs when there is a statistically significant difference between either the control and disturbed simulations of a climate model, and/or between the "normal" climate and a climatic fluctuation or trend caused by a "climatic forcing" of some kind.

climate/society interaction

Where feasible, people attempt to control their environment, yet complete control is elusive; the result is a continual interaction between social and natural systems. In some places, natural resources are well managed, and the environment is in harmony with human activity. In other cases, societies are confronted with depletion of natural resources, environmental degradation, and natural hazards. Social systems have generally evolved to adapt to or respond to natural forces in order to meet the needs of growing populations and the demands that accompany increased development and affluence.

The disruptive hazards and occasional disasters that indicate failure to achieve a sustainable nature–society interaction are often met by temporary and ad hoc responses. The climate impacts of the past two decades, a result of both increased social sensitivities and natural variation, suggest that to take the climate for granted, to respond to its impacts in a piecemeal fashion, or to neglect its fundamental rôle in social development and wellbeing, is to risk increased environmental problems in the future. While those problems may not emerge during the tenure of any single political leader or resource manager, they will affect the society that those leaders worked to shape. Both the scientific and political communities must ensure that greater efforts are made to incorporate climate considerations into natural resources management and into the way we structure human systems such as settlements, food production and energy distribution networks.

climate system

The climate system consists of the atmosphere, the hydrosphere (comprising the liquid water distributed on and beneath the surface of the Earth), the cryosphere (comprising the snow and ice on and beneath the surface of the Earth), the surface lithosphere (comprising the rock, sand, soil and sediment of the Earth's surface), and the biosphere (comprising the Earth's plant and animal life, as well as the activities of people). Principally under the influence of the solar radiation received by the Earth and its atmosphere, the climate system determines the climate of the Earth. Although climate relates essentially to the varying states of the atmosphere, the other parts of the climate system as described above have a significant rôle in the formation and development of climate(s) through their interaction with the atmosphere.

climate system monitoring

The addition of the word monitoring, by the Eleventh Session of the WMO Congress which met in Geneva in May 1991, to the name of the former World Climate Data Programme underscores the importance of this activity. A meeting of experts was held in August 1991 to begin the preparation of the WMO Climate System Monitoring Biennial Review for the period December 1989 through May 1991. The Review, the fourth in the series beginning in 1982, will encompass the entire global climate system and is co-sponsored by WMO and UNEP. The content will include *inter alia*, descriptions of global and regional temperature patterns, drought and desertification, the El Niño pattern and its global effects, snow and ice coverage, ocean temperatures and currents, and atmospheric trace-gas and aerosol concentration.

Climate System Monitoring Project (CSM)

This is a WMO project which provides WMO Members with information on large-scale climatic fluctuations and to facilitate the interpretation and dissemination of this information.

The aim of the project is to collect, synthesize, summarize and disseminate concise information on significant fluctuations or anomalies of the global climate system utilizing analyses and products from existing data and climate diagnostic centres (e.g. WMCs). Specific activities associated with this project include:

* Monthly CSM Bulletins: This bulletin contains global analyses of temperature and precipitation anomalies and statistics (by region and country) which indicate the persistence of hot/cold, wet/dry events, global analyses of sea-surface temperature, outgoing longwave radiation (an index of precipitation in the tropics), circulation anomalies, and drought-monitoring indices.
* Special Advisories: Through monitoring the global climate system, information is be provided on climatic events that affect a particular country or countries (in a region). This depends on the availability of relevant information and data-analysis centres to prepare the necessary outputs.
* Annual CSM Summaries: Summarized analyses will be prepared on the state of the climate system and significant large-scale anomalies.
* Biennial Scientific Reviews. These reviews provide diagnostic insights into significant anomalies, and possible cause–effect relationships and teleconnections.

climate variability

In the most general sense, the term climate variability denotes the inherent characteristic of climate which manifests itself in changes in the climate over time.

The degree or magnitude of climate variability can be described by the differences between the long-term statistics of the meteorological elements calculated for different periods. In this sense, the measure of climate variability is essentially the same as the measure of climate change.

The term climate variability is also used to describe deviations of climate statistics over a given period of time (such as a month, season, year) from the long-term statistics relating to the same calendar period. In this sense, the measure of climate variability is generally termed a climate anomaly.

climate/weather forecasting

The familiar weather forecast is similar to a climate forecast with the important exception that weather forecasts typically describe in some detail the expected weather conditions during the next few hours to few days, whereas climate forecasts describe in much more general terms the expected "climate" conditions over the next few weeks to few months. Another important distinction between weather and climate forecasts is that climate forecasts do not and in fact cannot predict that, for example, it will snow in New York on Christmas Day, whereas a weather forecast issued up to 5 to 10 days before Christmas Day could well do so, and moreover may be correct.

To make a climate forecast it is necessary to take into account all of the complex interactions and feedbacks between the different components of the climate system. This can be done through the use of numerical models which, so far as possible, include a description of all of the processes and interactions.

climatic analogue

A climate situation in the past which produced changes similar to that occurring in the present. Climatic analogues are used in making climatic projections.

climatic anomaly

The difference between the value of a climatic element at a given place and the mean value of that element/or departure from the normal value.

climatic classification

The division of the Earth's climates into a worldwide system of contiguous regions, each of which is defined by the relative homogeneity of the climatic elements. There are several global climatic classifications, including those originated by Köppen and Thornthwaite.

climatic cycles

Periodic rhythms in a long series of observations of climatic elements.

climatic element

Any one of the properties or conditions of the atmosphere (such as air temperature) which together specify the physical state of the weather or climate at a given place, for any particular moment or period of time.

climatic factors

Certain physical conditions (other than the climatic elements) which control the climate (latitude, height, distribution of land and sea, topography, ocean currents, etc.).

climatic fluctuation

A climatic inconstancy that consists of any form of systematic change, whether regular or irregular, except trends and discontinuities. Characterized by at least two maxima (or minima) and one minimum (or maximum), including those at the end points of the record.

climatic forcing see radiative forcing

climatic noise

It is highly desirable to distinguish the "human-induced" climate signal from that of the "undisturbed or natural" climate signal. To do so the "noise" or the known causes of climate change need to be identified. Climatic noise in this context includes carbon dioxide and other trace gas concentrations, solar radiance changes, stratospheric aerosol concentrations, tropospheric aerosol concentrations, El Niño/Southern Oscillation indices, surface albedo changes, and changes in North Atlantic deep water formation.

climatic normals

Period averages computed for a uniform period of 30 years (see also **climatological period averages**, and **climatological "standard normals"**)

climatic optimum

The period during which temperatures reached their highest levels within an interglacial stage of the Pleistocene period. The climatic optimum of the present interglacial occurred 4,000–7,000 years ago during which surface air temperatures were warmer than at present in nearly all regions of the world. In the Arctic region, the temperature rose many degrees, and in temperate regions the increase was 1.0–1.7°C. In this period, there was a great recession of glaciers and ice-sheets, and sea level was raised by about 3m.

climatic oscillation

A fluctuation in which the variable tends to move gradually and smoothly between successive maxima and minima.

climatic periodicity

A rhythm in which the time interval between successive maxima and minima is constant or very nearly constant through the record.

climatic rhythm

An oscillation, vacillation or cycle in which the successive maxima and minima occur at approximately equal intervals of time.

climatic trend

A climatic change characterized by a smooth, monotonic increase or decrease of the average value in the period of record. It is not restricted to a linear change with time, but characterized by only one maximum and one minimum at the end points of the record.

climatic vacillation
A fluctuation in which the climatic variable tends to dwell alternately around two or more average values, and to drift from one to the other at regular or irregular intervals.

climatic variation
A fluctuation or a component thereof, whose characteristic timescale is sufficiently long to result in an appreciable inconstancy of successive 30-year averages (normals) of the variable. It is often used to designate common natural variations from one year to the next, or changes from one decade to the next.

climatological network
All stations of a particular type (e.g. ordinary climatological stations), or stations participating in a special programme, irrespective of their type (e.g. sunshine network), in which the observed data are derived from official instruments with their exposure and observing practices conforming to prescribed standards.

climatological observation
Evaluation or measurement of one or several climatic elements.

climatological period averages
Averages of climatological data computed for any period of at least ten years starting on 1 January of a year ending with the digit 1 (i.e. 1931, 1941, 1951, etc.).

climatic record
Any record made of meteorological events in alphanumerical, graphical or map form.

climatological series
Homogeneous dataset of random variables, either discrete or continuous and selected from a single population, usually infinite in extent.

climatological "standard normals"
Averages of climatological data computed for the following consecutive periods of 30 years: 1 January 1901 to 31 December 1930, 1 January 1931 to 31 December 1960, 1 January 1961 to 31 December 1990.

climatological station
A station at which climatological observations are made.

climax

A mature plant community that is optimally adapted to existing environmental conditions and is therefore able to reproduce itself indefinitely. It is the final stage of development under a given set of conditions.

cloud

A visible aggregate of minute particles of liquid water and/or ice in suspension in the atmosphere. This aggregate may include larger particles of liquid water or ice, non-aqueous particles or solid particles, originating for example from industrial gases, smoke or dust.

cloud cover

The amount – usually measured in fractions of eighths or tenths – of the sky that is covered by clouds.

cloud feedback

The coupling between cloudiness and surface air temperature in which an increase in surface air temperature serves to change the extent, height and/or characteristics of the cloud cover such that the surface air temperature is further modified. It should be noted that increased cloud cover reduces the solar radiation reaching the Earth's surface, thereby lowering the surface temperature. An increase in middle and low-level clouds could increase the surface albedo, decrease the net downward solar radiation, decrease the surface air temperature, and cool the atmosphere–Earth–ocean system, resulting in a negative feedback. On the other hand, an increase in high-level clouds could increase the absorption of solar radiation, decrease the net outgoing radiation, increase surface air temperatures, and heat the atmosphere–Earth–ocean system, resulting in positive feedback.

cloud seeding

The introduction of particles of appropriate material (solid carbon dioxide, crystals of silver iodide, etc.) into a cloud, by aircraft or by ground generators, with a view to modifying the cloud structure and causing precipitation.

clouds and the Earth's climate

A major unsolved problem is that of reliably predicting the effect of clouds on the Earth's climate. Low-level clouds reduce the absorption of sunlight and tend to cool the Earth, while high clouds are essentially transparent to sunlight but trap infrared radiation from the Earth, thus contributing positively to the greenhouse warming of the surface. Not only is the cloud amount important to climate, but also cloud altitude and cloud optical properties. The matter is made more difficult by the lack of reliable global statistics of observed cloudiness, cloud properties or precipitation, which could serve to calibrate existing climate models. The WCRP has undertaken the International Satellite Cloud Climatology Project (ISCCP) to produce the required global cloud statistics.

CMM see **Commission for Marine Meteorology**

Coastal Zone Management Subgroup (CZMS)
The Coastal Zone Management Subgroup of IPCC WG III (Response Strategies).

COBIOTECH see **Scientific Committee for Biotechnology**

CODATA see **Committee on Data for Science and Technology**

coefficient of variation
A statistical parameter describing the change of a stochastic variable in time or space, expressed as the ratio of the standard deviation to the mean.

Commission for Aeronautical Meteorology (CAeM) of WMO
Responsible for matters relating to (among other things):
* Applications of meteorology to aviation, taking into account the relevant meteorological developments in both the scientific and practical fields;
* International standardization of methods, procedures and techniques employed or appropriate for employment in the application of meteorology to aeronautics and the provision of meteorological services to international air navigation, and the making, reporting and dissemination of meteorological observations from aircraft;
* Consideration of requirements for climatological data needed for aeronautical meteorological purposes.

Commission for Agricultural Meteorology (CAgM) of WMO
The Commission for Agricultural Meteorology (CAgM) of WMO is responsible for matters relating to (among other things):
* Applications of meteorology to agriculture cropping systems, forestry and agricultural land-use and livestock management, taking into account meteorological and agricultural developments in both the scientific and the practical fields;
* Development of agricultural meteorological services of WMO Members by the transfer of knowledge and methodology, and by providing advice: in particular (i) the most practical use of knowledge concerning weather and climate for agricultural purposes such as conservation of natural resources, land management, intensification of crop production, increase in the area of agricultural production, reduction of production costs, the improvement of agricultural products and the selection of improved varieties of plants and breeds

of animals that are better adapted to the climatological conditions and their variability; (ii) the combating of unfavourable influences of weather and climate on agriculture and animal husbandry, including weather-related pests and diseases; the protection of agricultural produce in storage or in transit against damage or deterioration due to the direct and indirect influences of weather and climate; (iii) the use of weather and agrometeorological forecasts and warnings for agricultural purposes; and (iv) the interactions between air pollution and vegetation and soil;
* Methods, procedures and techniques for the provision of meteorological services to agriculture including farmers and forestry and rangeland operators; and the formulation of data requirements for agricultural purposes.

Commission for Atmospheric Sciences (CAS) of WMO

Responsible for matters relating to (among other things):
* Research in meteorology and in related fields, and in particular:
* The development of research programmes in specific areas, primarily: weather prediction, including short-, medium-, and long-range; tropical meteorology;
* Research on climate, including climate variations, and as far as possible climate prediction, taking into account the special arrangements made for the World Climate Research Programme (WCRP);
* Weather modification; atmospheric chemistry and air-pollution meteorology, including studies of transport, transformation and deposition of air pollutants as lead technical commission in this field;
* Formulation of requirements for observations and for the storage, retrieval and exchange of data for research purposes;
* The co-ordination of the international aspects of such research with relevant scientific bodies;
* Standardization and tabulation of physical functions and constants used in atmospheric sciences; scientific evaluation of technical meteorological procedures such as forecast verification techniques; and, research techniques and methodology.

Commission for Basic Systems (CBS) of WMO

Responsible for matters relating to (among other things):
* Worldwide co-operation in the operation and further development of the World Weather Watch (WWW) system in the light of new requirements and technological developments;
* Development and application of systems and techniques for operational weather analysis and forecasting to meet user requirements;
* Observational systems, facilities and networks (land, sea, air and space) for basic meteorological purposes, and in particular all technical aspects of the Global Observing System (GOS) of the World Weather

Watch;
* Processing of basic data, regardless of whether the data refer to WWW stations or not, and the functions of appropriate data-processing centres in respect of processing, storage and retrieval of data for meteorological and related purposes, in particular the organization of the Global Data-Processing System (GDPS) of the WWW;
* Telecommunications networks and facilities to meet the basic requirements for operational, research and applications purposes, in particular the organization of the Global Telecommunication System (GTS) of the WWW;
* Operational procedures, schedules, arrangements to be developed for the international exchange of observational data and processed information for all purposes (including the World Climate Programme) and in particular through the GTS;
* Monitoring the operations of the World Weather Watch;
* International code forms and tables of specifications for basic processing purposes and for the various applied purposes;
* Formulation of system requirements to provide data and products to meet the requirements defined by the WMO Technical Commissions, the Regional Associations and other bodies taking into account new applications of meteorology;
* Definition of requirements for climatological data needed for general-purpose forecasts.

Commission for Climatology (CCI) of WMO

Responsible for matters relating (among other things) to:
* The study of climate (except for research on climate undertaken by the WMO Commission for Atmospheric Sciences) and its effects on human activities;
* The compilation and consolidation of general data requirements for all components of the WCP (in co-operation with other relevant bodies) as the Commission having the lead rôle in the WCDAP;
* Specification of requirements for climatological purposes and to meet the needs of users for instruments; observations; station networks; acquisition, quality control, inventories, exchange and archiving of data, proxy data and palaeoclimatic data;
* The development and improvement of application methodologies (in co-operation with other relevant WMO Commissions) as lead Commission in the WCAMP, in particular for the application of meteorological (especially climatological) information in the fields of energy, land-use and human settlements, engineering and building, human wellbeing (especially health and disease), tourism, industry, transportation (especially on land) and communications, economic and social planning;
* Statistical methods for describing and interpreting large sets of climatological data, assessment of the representativeness and general reliability of climatological observations and of the homogeneity of

climatological series;
* Studies of human effects on climate on local and regional scales, including local air-pollution climatology;
* The development and effective presentation to users of climatological information, in particular for the purposes of the WCAMP.

Commission for Hydrology (CHy) of WMO

Responsible for matters relating to (among other things):
* Activities in operational hydrology and the applications of meteorology and hydrology to water-resources problems;
* International standardization of methods, procedures, techniques and terminology for studies of the water balance, the global hydrological cycle and hydrological forecasting, and meteorological and hydrological aspects of design of systems for water management and control;
* Reliability and homogeneity of hydrological and related meteorological observations;
* Standardization of the form for recording and establishing requirements for the exchange of hydrological observations and for their processing;
* Standardization of methods of computation of hydrological data for research and publication (means, ranges, frequencies, etc.).

Commission for Instruments and Methods of Observations (CIMO) of WMO

Responsible for matters relating to (among other things):
* International standardization or compatibility of meteorological instruments and related measurements and observations;
* Support to other WMO programmes and bodies, in particular by specifying co-ordinated requirements for meteorological measurements methods, when requested by other technical commissions;
* Promotion of research and development for automatic observations and of meteorological instrumentation including inexpensive and sturdy instruments for use in developing countries.

Commission for Marine Meteorology (CMM) of WMO

Responsible for matters relating to (among other things):
* Applications of meteorology and relevant parts of physical oceanography to marine activities in open-sea, offshore and coastal areas, taking into account new developments in research and operations;
* Meteorological and related oceanographic aspects of international programmes of ocean investigations and explorations;
* Marine meteorological services carried out by Members which at the same time constitute part of the services of the joint WMO/IOC Integrated Global Ocean Services System. These responsibilities

include in particular:
* Recruitment of ships and the training of marine observers including organization of the transmission of ship observations to shore;
* Observations of ocean-surface conditions including temperature, currents, waves and sea ice and applications of such data;
* Services to various marine activities normally carried out by national meteorological authorities including specialized services for offshore industry and fisheries;
* Standardization of methods, procedures and techniques for marine meteorological observations and, in co-operation with other interested organizations, related physical oceanographic observations;
* Standardization of observations of ocean waves, sea ice and sea temperature and studies of these elements;
* Preparation of climatological summaries and other climatological information for marine purposes.

Committee on Climatic Changes and the Ocean (CCCO)
The Committee on Climatic Changes and the Ocean (CCCO) was established by SCOR and held its first meeting in 1979. It is now co-sponsored by IOC. The main function of the Committee is to improve our understanding of the ocean's rôle in climate change and variability and to identify the most important climatologically significant processes and the means for their incorporation into mathematical models. CCCO co-operates closely with the ICSU–WMO Joint Scientific Committee for the WCRP.

Committee on Data for Science and Technology (CODATA)
The Committee was set up in 1966 by the 11th General Assembly of ICSU held in Bombay. The Committee held its first meeting in Paris in June 1966. CODATA is concerned with all types of quantitative data resulting from experimental measurements or observations in the physical, biological, geological and astronomical sciences. Particular emphasis is given to data-management problems common to different scientific disciplines and to data used outside the field in which they were generated. The general objectives are the improvement of the quality and accessibility of data, as well as the methods by which data are acquired, managed and analyzed; the facilitation of international co-operation among those collecting, organizing, and using data; and the promotion of an increased awareness in the scientific and technical community of the importance of these activities.

Committee on Science and Technology in Developing Countries (COSTED)
This Committee was set up in 1966 by the 11th General Assembly of ICSU held in Bombay, for the encouragement of science and technology in developing countries. To achieve this purpose, the Committee has the following objectives: to co-ordinate and encourage efforts by the

International Scientific Unions to assist the developing countries; to work with the Special and Scientific Committees of ICSU in order to facilitate the greatest possible participation in their programmes by scientists in the developing countries; to foster affiliations with COSTED of national or regional committees, which could identify scientific and technical problems related to developing countries and recommend programmes and other activities for COSTED; to provide liaison and advisory services, when requested, to international and regional scientific organizations; to consider the methodology of using science and technology to assist the developing countries; and to undertake other activities and programmes designed to assist developing countries through the use of science and technology.

Committee on Space Research (COSPAR)

COSPAR was set up by the 8th General Assembly of ICSU in Washington DC, in 1958 to continue the co-operative programmes of rocket and satellite research undertaken during the International Geophysical Year. COSPAR is an interdisciplinary scientific body concerned with the progress on an international scale of all kinds of scientific investigations carried out with space vehicles, rockets and balloons. The membership of COSPAR is composed of representatives of national academies of science or the equivalent and representatives of the International Scientific Unions.

Committee on Water Research (COWAR)

The Committee on Water Research (COWAR) was created by ICSU in 1964. The rôle of COWAR is to promote necessary contacts between international non-governmental water-orientated organizations, so that their activities pertaining to water research will complement each other for the common benefit of all the organizations and of the people they ultimately serve.

commodity agreements

Agreements concluded between commodity-producing and commodity-importing countries with the aim of stabilizing the export prices for the producing countries. This is done primarily by means of intervention prices, buffer stocks and export quotas. Such agreements ensure that consuming countries have access to a steady supply of commodities at reasonable prices.

Commonwealth Expert Group on Climate Change

In 1988, the Secretary-General of the British Commonwealth established a small expert group to consider the implications of climate change to all countries, but with particular reference to Commonwealth countries. The report of the Expert Group was published in September 1989. In the foreword to the report the Commonwealth Secretary-General states:

This group breaks new ground in several ways. First, the membership was predominantly from developing countries. The development perspective – the concern for the poor – is kept well in the foreground. The report makes it plain that the world's poor could be the main victims of climate change but that this must not be allowed to occur; it concludes that a global strategy for controlling global emissions must permit rapid economic growth in developing countries. Secondly, the report is practical and businesslike in suggesting how some planned adaptation to climate change can take place. It sounds an alarm, but is not alarmist. It sketches out how a vulnerable small island state, or those responsible for farming in drought-prone regions, can prepare for, and mitigate, the worst consequences of climate change. And it provides a detailed guide to the kind of data-collection effort which is needed for governments to monitor and analyze the changes taking place, so as to help individuals adjust to known facts.

In the near future, a major international initiative will be needed to establish global responsibilities for preventing unmanageable rates of increase of global temperature. This requires both technical preparation of the highest order and a great deal of political will. This report has contributed substantially to the former and helps to create a basis for the latter through a clearly formulated Commonwealth Plan of Action.

community

In ecology, a unit in nature comprising populations of organisms of different species living in the same place at the same time.

compliance date

The date upon which a pollutant source is required to meet applicable pollution control requirements.

confidence interval

A statistical parameter describing the interval which includes the true value with a prescribed probability and which is a function of the statistics of the sample.

confidence level

A statistical parameter describing the probability that the confidence interval includes the true value.

condensation

The conversion of water vapour into water droplets in the form of fog, clouds or dew.

continental drift

An hypothesis proposed around 1910 to describe the movements of continental masses over the surface of the Earth. A major theorist of continental drift, and certainly the one who gave the hypothesis scientific plausibility, was Alfred Wegener (1880–1930). His work was based on qualitative data, but this has been vindicated in recent years by plate tectonics theory, which has provided geologists with a viable mechanism to account for continental movements. The evidence for the existence of continental drift comes from palaeomagnetism, the distribution of orogenic belts, faunas and climatic belts, and the morphological fit along the edges of the continental shelves.

Convention of the World Meteorological Organization

The Convention of the World Meteorological Organization – among other things – states (that) with a view to co-ordinating, standardizing, and improving world meteorological activities, and to encouraging an efficient exchange of meteorological information between countries in the aid of human activities, the contracting states agree that the purposes of the Organization shall be:

* to facilitate worldwide co-operation in the establishment of networks of stations for the making of meteorological observations, as well as hydrological and other geophysical observations related to meteor-ology, and to promote the establishment and maintenance of meteorological centres charged with the provision of meteorological services;
* to promote the establishment and maintenance of systems for the rapid exchange of weather information;
* to promote standardization of meteorological observations and to ensure the uniform publication of observations and statistics;
* to further the application of meteorology to aviation, shipping, agriculture, and other human activities; and
* to encourage research and training in meteorology and to assist in co-ordinating the international aspects of such research and training.

convention treaty

Generally, "convention" in international law is used interchangeably with "treaty"; both imply a consensus document. A treaty, however, usually contains substantive commitments. The term "framework convention" is used to denote a document that outlines general commitments, much like the Vienna Convention did for ozone-depleting substances. Protocols are designed to address specific portions of a wider commitment. For example, in any framework convention on climate change, each greenhouse gas would likely be addressed under a separate protocol.

cooling degree-day

Defined as the number of degrees by which the average daily temperature exceeds a threshold or base temperature such as 25°C or 30°C. The number of cooling degree-days in a season is the summation of the cooling degree-days for all days, and it provides an index of the air conditioning (or cooling) requirements of a particular area.

cooling lake

Any manmade water impoundment which impedes the flow of a navigable stream and which is used to remove waste heat from heated condenser-water prior to recirculating the water to the main condenser.

cooling tower

A device that aids in heat removal from water used as a coolant in electric-power generating plants.

Coriolis effect

The deflection of a body in motion (such as currents of air and water) due to the Earth's rotation; such deflection is to the right in the Northern Hemisphere and to the left in the Southern Hemisphere.

COSPAR see **Committee on Space Research**

cost–benefit analysis

The assessment of the total costs and benefits to society and ecosystems associated with a particular activity or event, including indirect and intangible costs and benefits. This term is commonly used to describe the process of assessing the practicality of a specific action or strategy, or alternatively, the net effects of an event, usually in quantitative, economic terms. However, qualitative factors related to social, environmental or other externalized costs and benefits are very often inadequately dealt with in such analyses, and more emphasis needs to be placed on developing methods to assess "total" or "true" costs and benefits of actions or events.

costs of responding to climate change

If current perceptions about the range of possible environmental and socio-economic effects are substantially correct, then the costs incurred by doing nothing about climate change could be extremely large. However, strategies for either adapting to climate change or limiting it by controlling greenhouse gas emissions, or both, will also involve high costs to global society. Clearly, it would be preferable to be more certain about the magnitude and rate of onset of global warming, and about its environmental and socio-economic effects, before taking expensive adaptation and/or limitation actions. But time is not on our side and, at least for policy-making purposes, there is an urgent need for detailed comparisons of the costs of various strategies.

Some of the items that need to be included in any programme for comparing the costs of different strategies can be made by considering four scenarios: Business-as-Usual, Moderate Effort and Concerted Effort, and Surprise. The different scenarios reflect the level of effort and ambition of policies that can be undertaken specifically to deal with climate change induced by greenhouse gases. The Business as Usual scenario implies that no policies explicitly directed at greenhouse gas limitation are undertaken, whereas Moderate Effort and Concerted Effort reflect the level of effort devoted to such items as energy policy, reforestation, and greenhouse gas reduction strategies. The Surprise scenario differs from the other three, since it could occur in any one of the other scenarios, although it is perhaps less likely to occur in the case of Concerted Effort than that of Business-as-Usual. It is intended to highlight the consequences of an unpredicted surprise event, such as an abrupt change of climate as a result of an unpredicted change of the oceanic circulation.

The response limitation refers to specific adaptation steps taken before effects occur, while Forced Adaptation occurs in response to physical and biological impacts. Residual costs are those for which adaptation steps are not, or cannot be, undertaken, and will largely be external costs involving such things as the global commons, unmanaged ecosystems and human suffering. It should be emphasized that adaptation strategies will have to be replaced ultimately by limitation strategies, since a significant warming would sooner or later become intolerable, no matter how much is spent on adaptation. Limitation strategies on the other hand cannot totally limit emissions, and some adaptation will be required, especially as a result of the emissions that have already occurred.

coupled ocean–atmosphere general circulation models

The slowly changing response of climate to a gradual increase in greenhouse gas concentrations can be modelled rigorously only using a coupled ocean–atmosphere general circulation model with full ocean dynamics. This has now been done by a few researchers using coarse-resolution models out to 100 years. Their results according to the IPCC Scientific Assessment Report show that the results are generally consistent with our understanding of the present circulation in the ocean, as evidenced by geochemical and other tracers. However, available computer power is still a serious limitation on model capability, and existing observational data are inadequate to resolve basic issues about the relative rôles of various mixing processes, thus affecting the confidence level that can be applied to these simulations.

· Based on the IPCC Business-as-Usual scenarios, the energy-balance upwelling diffusion model with best judgement parameters yields estimates of global warming from pre-industrial times (taken to be 1765) to the year 2030 of between 1.3°C and 2.8°C, with a best estimate of 2.0°C. This corresponds to a predicted rise from 1990 to 2030 of 0.7°C to 1.5°C with a best estimate of 1.1°C. The model also estimates the temperature rise from pre-industrial times to the year 2070 to be between 2.2°C and 4.8°C with a best estimate of 3.3°C. This corresponds to a predicted rise from 1990 to 2070 of 1.6°C to 3.5°C, with a best estimate of 2.4°C.

COWAR see **Committee on Water Research**

crop calendar

A list of the standard crops of a region in the form of a calendar giving the dates of sowing and the agricultural operations, and the various stages of their growth in years of normal weather.

crop/climate calendars

Specific and relative climate sensitivities of certain crops in an area might be discerned by comparing standard crop calendars (timetables of the phenological stages achieved during a normal year) to the risk of adverse weather and climate conditions at each period. It is well known, for example, that young plants are more susceptible to cold, that maize is particularly drought- and heat-sensitive during its flowering period, and that grain is sensitive to excessive moisture during the "dry-down" period, when it hardens just before or during harvest. Because sensitivities vary from place to place and among crops, a crop-sensitivity calendar should be devised for each major crop and agricultural production zone of concern.

crop growth and carbon dioxide

There are widely differing estimates of the net effects on global agriculture of increased carbon dioxide and global warming. Some assessments have predicted deteriorating conditions in agriculture, others have been more optimistic. Given the uncertainties involved and the different ways climate affects agriculture, this is not surprising. A positive influence from global warming arises from the effect of the higher levels of atmospheric CO_2 on photosynthesis and plant growth. Laboratory experimental results suggest that a doubling of carbon dioxide concentration could cause a 10–50% increase in the yields of a wide range of so-called C3 crops, such as wheat, rice, potato, barley, cassava, oilseeds, beet sugar and most fruits and vegetables, that collectively contribute about 80% of world food supplies. A smaller increase of 0–10% could come from a second group (the so-called C4 crops) including maize, sorghum, millet and cane sugar; and none from a minor third group (CAM) including pineapples. Laboratory experiments do not, however, reflect the complex conditions observed in nature where, for example, the growth of pests and weeds would compete with that of commercial crops. Also, more fertilizer may be required to achieve the potential increase in yield. While the results have to be treated with caution, they do nonetheless suggest potential gains which could be substantial for some commodities.

crop weather modelling approaches

The most common method for extrapolating agricultural impacts of future climate conditions is to apply existing crop-yield models to a predicted set of climate conditions. The models fall into two categories:

empirical–statistical and dynamic-simulation. Empirical–statistical models typically are based on multiple-regression equations relating climate conditions (perhaps for specific phenological periods) to yields over an historical time period with matched climate and crop-yield datasets. Dynamic-simulation models (also called physical or deterministic models) use mathematical formulae to simulate the physical processes by which environmental, chemical and biological factors interact to produce crop yields. By varying the environmental factors in accordance with potential future climate conditions, the models simulate future impacts on yields.

Statistical and physical models have been developed for most of the world's major crops but are not easily transferred from one field or laboratory setting to another. Thus, impact assessors should try to use proven, regionally developed models. Most models of this sort are developed for short-term yield forecasting or for research purposes, but they can be modified to project future effects.

Few attempts have been made to model future crop yields, and most existing work focuses on wheat and maize grown in Europe or North America. There are three critical weaknesses in modelling approaches: the climate projections themselves may be unrealistic; the models assume a level of technology which might change in the future; and the models may be particularly unreliable when used to project impacts of climates substantially altered from those under which they were developed. Projections of future climate used to extrapolate impacts might come from atmospheric models, historical analyses, or they may simply represent "reasonable increments" of climate change.

crop-yield analysis

In agriculture, the simplest impact assessment is to compare yields during climate fluctuations to those of more normal years. Typically, "normal" yields in, say, kg/ha are defined as an average over several years. Unfortunately, good historical records of crop yields are not available for several of the world's crop-producing regions. While some areas, such as North America and Europe, have reliable records for the past 50 years or more, many regions lack even a decade of consistent data. However, where no direct evidence exists for agricultural impacts, it is still possible to infer yield impacts by analogy from the better-documented cases.

Statistical or physical crop models can be applied where crop-yield statistics extend over several decades. In essence, crop–climate models express observed crop impacts as mathematical relationships between selected climate and other factors (e.g. soils and management) and yield. Such models are typically used to predict yields associated with projected weather and climate conditions, but they can be useful in empirical impact studies in two ways. First, verified yield models can be used to test the reasonableness of claims that a certain climate change was responsible for observed yield declines. However, it should be noted that statistical models tend to underestimate the impacts of extreme climate events (e.g. conditions not experienced during the data period for which the model was developed). Secondly, models can be used to extrapolate impacts back to periods for which yields data are lacking, thus aiding the interpretation of historical climate impacts. However, the researcher must remember that most models assume either a fixed or a sliding level of agricultural technology, and earlier technology may be quite different from that for which the model is calibrated.

crop-yield sensitivity analysis

One of the best indicators of climate sensitivity is crop-yield variability. Yield data are most often reported by farmers themselves and occasionally verified with spot-checks by agricultural experts. Yield might be reported as total weight or production, in which case the yield per unit area can be calculated if crop acreage is known. It is important in such cases to ascertain whether production statistics represent the yield per planted, or per harvested, hectare. Poor crop conditions can lead to significant crop abandonment. Because abandoned fields are probably the poorest lands – those most sensitive to climate impacts – yield information that does not consider abandonment can lead to inflated estimates.

New cultivars are usually introduced to increase yields, but at the cost of increased yield variability. Modern, high-yielding cultivars, with their additional input requirements, may actually be less resilient to climate fluctuations, but, in theory, the broader socio-economic systems to which they belong, with associated food storage and expanded opportunities for off-farm income, make a strategy of greater average yield (with occasional very poor yields) socially acceptable. Such a risky approach would not, however, be acceptable to self-provisioning farmers or in regions where locally produced food provides a significant proportion of nutrition.

Where at least a decade of crop-yield data are available, a good idea of overall climate sensitivity can be obtained simply by examining the record. Except where short-term swings in inputs are common (due to fluctuations in the price of fertilizers, for example), interannual yield variability is primarily caused by climate, pests or disease. If the researcher can rule out pests and disease, then changes in yield from one year to the next are undoubtedly climate-induced. This is especially true wherever a single climate element (e.g. growing-season length) is the chief constraint on crop production. The great majority of interannual crop-yield variability in dryland crop areas is attributable to rainfall fluctuation, though rainfall variability can also affect yields in moist regions. Yield variability in high latitudes is usually caused by temperature variations.

cross-dating

The matching of tree-ring width patterns and other properties among the trees and fragments of wood from a particular area. This enables the year in which each ring was formed in living trees and recent stumps to be determined accurately, the presence of false rings or the absence of rings in individual specimens being made apparent. By matching ring series from living specimens with those from older (e.g. constructional) timbers, the chronology may be extended backwards in time.

CRU

The climatic research unit concerned with the understanding of climate, and past, present and future fluctuations, located at the University of East Anglia, UK.

cryosphere

The portion of the climate system consisting of the world's ice masses and snow deposits, which includes the continental ice sheets, mountain glaciers, sea ice, surface snow cover, and lake and river ice. Changes in snow cover on the land surfaces are largely seasonal and closely tied to the mechanics of atmospheric circulation. The glaciers and ice sheets are closely related to the global hydrological cycle and to variations of sea level, and change in volume and extent over periods ranging from hundreds to millions of years.

CSERGE see **Centre for Social and Economic Research on the Global Environment**

CSM see **Climate System Monitoring Project**

D

D-layer

A part of the ionosphere, situated at a height between about 65km and 80km, which is mainly responsible for the absorption of radio energy reflected from higher levels, but from which radio waves of very low frequency may also be reflected.

daily (diurnal) range of temperatures

Temperature range in the course of a continuous time interval of 24 hours.

daily maximum/minimum temperatures

Maximum/minimum temperature in the course of a continuous time interval of 24 hours.

DARE see **Data Rescue Programme**

data processing
The handling of data until they are in a form ready to be used for a specific purpose.

Data Rescue Programme (DARE)
DARE is a WMO programme, initiated in 1986, and designed to save rapidly deteriorating written meteorological records in some developing countries, by microfilming the information, then entering it into computers for quality control so as to provide easier access to long-term records for studying trends in climate and extreme events. DARE is being carried out in developing countries through the generous support of national donors and international agencies, and in particular the International Data Co-ordination Centre in Brussels, Belgium.. The project is now fully implemented in 15 countries in Africa, resulting in microfilming of more than one million observational records. UNEP has provided funds to purchase microfilming and microfiche-reading equipment in support of the DARE project.

DBCP see Drifting Buoy Co-operation Panel

deforestation
Those forestry practices or processes that result in a long-term land-use change from forest to agriculture or human settlements or other non-forest uses.

deforestation and reforestation (IPCC WG I: Policymaker's Summary)
People have been deforesting the Earth for millennia. Until the early part of this century, this was mainly in temperate regions; more recently it has been concentrated in the tropics. Deforestation has several potential impacts on climate: through the carbon and nitrogen cycles (where it can lead to changes in atmospheric carbon dioxide concentrations), through the change in reflectivity of terrain when forests are cleared, through its effect on the hydrological cycle (precipitation, evaporation and runoff), and surface roughness, and thus atmospheric circulation which can produce remote effects on climate.

It is estimated that each year about 2Gt of carbon (GtC) is released to the atmosphere due to tropical deforestation. The rate of forest clearing is difficult to estimate; probably until the mid-20th century, temperate deforestation and the loss of organic matter from soils was a more important contributor to atmospheric carbon dioxide than was the burning of fossil fuels. Since then, fossil fuels have become dominant; one estimate is that around 1980, 1.6GtC was being released annually from the clearing of tropical forests, compared with about 5GtC from the burning of fossil fuels. If all the tropical forests were removed, the input is variously estimated at from 150GtC to 240GtC; this would increase atmospheric carbon dioxide by 35–60ppmv.

To analyze the effect of reforestation we assume that 10 million hectares of forests are planted each year for a period of 40 years, i.e. 4 million km² would then have been planted by 2030, at which time 1GtC would be absorbed annually until these forests reach maturity. This would happen in 40–100 years for most forests. The above scenario implies an accumulated uptake of about 20GtC by the year 2030 and up to 80GtC after 100 years. This accumulation of carbon in forests is equivalent to some 5–10% of the emission due to fossil-fuel burning in the business-as-usual scenario.

Deforestation can also alter climate directly by increasing reflectivity and decreasing evapotranspiration. Experiments with climate models predict that replacing all the forests of the Amazon Basin by grassland would reduce the rainfall over the basin by about 20% and increase mean temperature by several degrees.

degree-days

A convenient temperature index commonly used in agriculture, heating and air-conditioning. It is the algebraic difference, expressed in degrees, between the mean temperature of a given day and a reference temperature. Growing degree-days are used in agriculture, cooling degree-days in air conditioning, and heating degree-days in assessing heating requirements. (See also **cooling degree-days, growing degree-days, heating degree-days**)

dendrochronology

Dendrochronology is the science of using tree rings to date, among other things, the climate in the past. In particular, the annual growth rings of trees growing may yield information on year-to-year climate fluctuations.

The approximate age of a temperate forest tree can be determined by counting the annual growth rings in the lower part of the trunk. The widths of these annual rings signify favourable growing conditions, absence of diseases and pests, and favourable climatic conditions, while narrow rings indicate unfavourable growing conditions or climate. The most sensitive (variability in ring widths) tree-ring chronologies come from trees whose growth has been limited in some way by climatic or environmental factors. Tree rings record responses to a wider range of climatic variables over a larger part of the Earth than any other type of annually dated proxy record.

denudation

The erosion by rain, frost, wind or water of the solid matter of the Earth. Denudation often implies the removal of the soil down to the bedrock.

depression see **low**

desert

A region where the precipitation is insufficient to support any except xerophilous vegetation; a region of extreme aridity.

desertification

The progressive destruction or degradation of vegetative cover, especially in arid or semi-arid regions bordering existing deserts. Overgrazing of rangelands, large-scale cutting of forests, drought and burning of extensive areas all serve to destroy or degrade the land cover.

In many areas the deserts are spreading, with a speed of about one or more kilometres per year (for example, the Sahara Desert was advancing into the Sahel – consisting of Mauritania, Senegal, Mali, Upper Volta, Niger and Chad – during the 1968–73 drought at a rate of 50km per year), depending on the density of the population and the consequence of grazing animals (especially goats). However, some desertification is the result of the interaction of naturally recurring drought along with unwise land-use practices.

The climatic impacts of desertification include an increased surface albedo, leading to decreased precipitation and decreased soil moisture, which in turn leads to less vegetative cover. Increased atmospheric dust loading can also lead to decreased monsoon rainfall and greater wind erosion and/or atmospheric pollution.

detection of the human-induced greenhouse effect on global mean temperatures

According to the IPCC Scientific Assessment Report, global-mean temperature has increased by 0.3–0.6°C over the past 100 years. The magnitude of this warming is broadly consistent with the theoretical predictions of climate models, but it is not yet established that the observed warming (or part of it) is attributable to the enhanced green-house effect. This is the detection issue. If the sole cause of the warming were human-induced greenhouse effect, then the implied climate sensi-tivity would be near the lower end of the range of model predictions.

Natural variability of the climate system could be as large as the changes observed to date, but there are insufficient data to be able to estimate its magnitude or its sign. If a significant fraction of the observed warming were due to natural variability, then the implied climate sensi-tivity would be even lower than model predictions. However, it is possible that a larger greenhouse warming has been offset partially by natural variability and other factors, in which case the climate sensitivity could be at the high end of model predictions.

deterministic model

A mathematical model in which all relationships are fixed and the concept of probability does not enter; that is, a given input produces one exact prediction as output. The opposite is a stochastic model.

deterministic system

A system is said to be deterministic if its response at any time due to a given input is uniquely determined.

dew

The deposit of water drops on objects at or near the ground, produced by the condensation of water vapour from the surrounding air.

dew point

The temperature to which air must be cooled in order for it to become saturated with water vapour.

diagnostic studies

The analysis of data to understand the relationships between components of the climate system. They provide the link between research and operational climate monitoring, especially through the provision of statistical information on circulation variables.

diffuse radiation

Radiant energy coming from different directions in contrast to direct radiation. Examples of diffuse radiation include radiation scattered and reflected by water vapour, clouds, other atmospheric gases and aerosols.

direct solar radiation

That portion of the radiant energy received directly from the Sun, as distinguished from diffuse sky radiation, effective terrestrial radiation, or radiation from any other source.

delivered or end-point energy

The energy available to the final consumer. As a rule, this includes most secondary energy sources, e.g. coal, petroleum and gas products, electricity and district heating, but also directly usable primary-energy sources such as natural gas. Losses involved in the conversion of primary energy into delivered energy, together with the non-used portion of primary energy, can be quite high and of the order of 20–40%.

divergence

The divergence of the flux of a quantity (e.g. radiation or momentum) is an expression of the time rate of depletion of the quantity per unit volume. Negative divergence is termed "convergence" and it relates to the rate of accumulation. In meteorology, divergence (or convergence) is used mostly in relation to the velocity vector and so refers to the flux of air particles themselves. The "divergence of velocity" is a three-dimensional property which expresses the time rate of expansion of the air per unit volume.

Dobson unit
A unit for measuring the total amount of ozone in a vertical column above the Earth's surface. One hundred Dobson units are calculated on the basis of an air layer 1mm thick at an atmospheric pressure of 1013hPa and at a temperature of 298°K.

doldrums
A zone of calm or light variable winds, in the lower atmospheric layers, situated near the thermal equator.

doomsday syndrome
One of the major factors in the emergence of climate change as a political issue has been the human fascination with predictions of impending doom. Some people would say that, as doomsday scenarios go, the greenhouse effect is more scientifically credible than most, and this credibility is increased through the almost irresistible temptation to view events such as the disastrous 1988 North American drought as the foretaste of things to come. In a column under the title "Which Apocalypse Now?" in a recent issue of the *Australian Physicist*, the President of the Australian Institute of Physics explains, *inter alia*, the rise of the greenhouse effect as a substitute for the threat of global nuclear war, and the public confusion between the greenhouse effect and the hole in the ozone layer in terms of his three laws of apocalyptics: (1) people need an apocalypse, (2) you can't have more than one apocalypse at a time, (3) apocalypses sell newspapers and take your mind off the economy, but they threaten your scientific credibility.

There is certainly sufficient experience already in many countries to make it clear that it will be extremely difficult to maintain a balanced public perception of the level of threat posed by greenhouse-induced climate change, both directly and through its potential for the generation of conflict within and between nations. Possibly even more worrying is the prospect of backlash or apathy when the threatened Armageddon does not arrive within reasonable time, or when a new apocalypse arrives. Moreover, some would argue that there has been too little critical evaluation so far of the relative advantages and disadvantages to either individual nations or the global community of a warmer world.

downwelling
The accumulation and sinking of warm surface waters, especially along a coastline. A change of airflow of the atmosphere can result in the sinking or downwelling of warm surface water. The resulting reduced near-surface nutrient supply affects the oceanic productivity and meteorological conditions of the coastal regions in the downwelling area.

Drifting Buoy Co-operation Panel (DBCP)
The Drifting Buoy Co-operation Panel (DBCP), established jointly by IOC

and WMO, provides a mechanism for co-operation in all aspects of drifting buoy operations. Its objectives are to encourage the optimum use of any drifting buoy deployments being undertaken worldwide and increase the amount of drifting buoy data available to meet the objectives of major WMO and IOC programmes. The Panel also encourages and supports the establishment of action groups in particular programmes or regional applications to effect the desired co-operation in drifting buoy activities. The DBCP maintains a summary of drifting buoy data requirements, a catalogue of existing drifting buoy programmes, and a list of focal points for national contributions. It also identifies sources of data not currently reported on the GTS and is implementing a real-time quality control system.

drifting buoys

There are presently approximately 600 drifting buoys operated by 11 countries (Australia, Canada, Germany, France, Japan, Netherlands, New Zealand, Norway, South Africa, United Kingdom, USA). All of the buoys use the Argos communication and data-processing systems, with 35–50% transmitting data in real-time via the WMO Global Telecommunication System (GTS). About 10% of the buoys have no sensors and are used as tracers only. Most of the other buoys measure at least air pressure and/or sea-surface temperature. Drifting buoy activities are managed by the IOC–WMO Drifting Buoy Co-operation Panel.

drizzle

Fairly uniform precipitation composed exclusively of fine drops of water (diameter less than 0.5mm) which are very close to one another.

drought

A period of abnormally dry weather sufficiently prolonged for the lack of water to cause a serious hydrological imbalance (i.e. crop damage, water-supply shortage, etc.) in the affected area. Drought severity depends upon other things on the degree of moisture deficiency, the duration of the drought and the size of the affected area.

drought impact pathways

The human consequences of a severe drought depend largely on the ways in which the effects filter through the socio-economic and political fabric of a society. For example, reduced agricultural production or changes in hydroelectric output are not the ultimate concerns, but rather the degree to which people are affected in terms of income, health and community stability. That is, a reduced wheat yield does not – except in certain critical circumstances – directly affect a society, but a change in the price for wheat or an inferior substitute does.

A framework for tracing the impacts of drought occurrence needs to show the pathways that drought impacts could take, spanning local through to global spatial scales, and three major systems that could be affected (agricultural, economic, social). The initial disturbance originates from a meteorological event becomes an agricultural drought – as distinct from a meteorological drought – when agricultural production falls below a perceived threshold. The agricultural drought then translates into a drought impact when the stress is detected in the economic, social and political sectors. The degree to which the initial climate event is transformed into stress is influenced by several factors, including market prices, government policies, farm stability, and the degree to which the drought is seen as a local, regional or global problem.

drought index
A computed value which is related to some of the cumulative effects of a prolonged and abnormal moisture deficiency.

Drought Watch
The Australian Bureau of Meteorology has maintained a regular "Drought Watch" system for Australia since 1965. Once the unusually dry period begins to develop, the Bureau of Meteorology issues monthly "Drought Statements" until the rainfall deficiencies are removed. These statements, which include maps, tables of rainfall information, and a brief description of the main features of the rainfall patterns, are intended to assist government, business and the rural community in maintaining an up-to-date assessment of the situation and to provide an early indication of the need for contingency action or drought relief.

dumping
A term used in international trade for the unloading of large quantities of a particular product in another country at a price lower than the usual market price or even below costs. The most important form of long-term dumping, in which products can sometimes be exported at a loss, is that of state-subsidized exports. International organizations such as GATT or the EC include provisions to prevent dumping between member countries and to protect against dumping practices by third countries.

dust
Solid materials suspended in the atmosphere in the form of small irregular particles, many of which are microscopic in size. Dust is due to many natural and artificial sources such as volcanic eruptions, salt spray from the seas, blowing solid particles, plant pollen and bacteria, smoke and ashes of forest fires, and industrial combustion processes, etc.

Dust Bowl
A name given early in 1935 to the region in the south-central United States which at that time was afflicted by a very severe drought and dust storms. It included parts of five states: Colorado, Kansas, New Mexico, Texas and Oklahoma. The "Dust Bowl" resulted from a long period of

deficient rainfall combined with loosening of the soil by destruction of the natural vegetation.

dust devil
A small, vigorous, well developed whirlwind made visible by the dust, sand or other debris picked up from the ground.

dust veil index
A scale developed by H. H. Lamb which ranks volcanic dust veils in terms of the mass of ejected material, and the duration and maximum extent of spread of the veil. The Krakatoa eruption (1883) is ranked as 1,000, while the eruption of Mount Agung (1963) had an index of 800.

dynamic climatology
The study of observed elements (or derived parameters) of the atmosphere, particularly in relation to the physical and dynamic explanation, or interpretation, of either the contemporary climate patterns with their anomalous fluctuations or the long-term climate changes or trends.

E

Earth
The Earth is a spheroid which is somewhat flattened at the Poles relative to the Equator because of rotation about its axis – the so-called "equatorial bulge". The Earth's interior comprises a thin outer "crust" of rocks which is part of a solid "mantle" extending to a depth of about 3,000km; below this is probably a metallic fluid "outer core" of thickness 2,200km, and near the Earth's centre an "inner core" of radius 1,300km which is probably solid. The estimated temperature at the Earth's centre is in the range 2,000–10,000°K. The respective portions of surface covered by water and land on the Earth are: in the Northern Hemisphere 60.7% and 39.3%; in the Southern Hemisphere 80.9% and 19.1%; for the Earth as a whole 70.8% and 29.2%.

Earth Charter
The Earth Charter will embody basic principles which must govern the economic and environmental behaviour of peoples and nations to ensure "our common future". The Charter is expected to be produced at the United Nations Conference on Environment and Development (UNCED) held in Brazil in June 1992.

Earth's orbital variations

In the 1930s, M. Milankovitch computed periodicities of 96,000 years in the elliptical shape of the Earth's orbit (that is, the orbit becomes alternately more circular or more elliptical), 40,000 years in the tilt of the Earth (at present about 23.5°N or S), and 21,000 years in the relation of the seasons to the distance of the Earth from the Sun (that is, the Earth is at present nearest to the Sun in the Northern Hemisphere winter, but in 10,000 years time it will be farthest at that season.) These orbital changes affect the amount of solar radiation available at a given latitude in a given season, and evidence on past ice ages, particularly from cores drilled in seabed sediments, has shown oscillations in climate on about these timescales. However, the changes are too slow to be capable of influencing the Earth's climate within the timescale of human generations. (See also **astronomical theory of climate change** and **Milankovitch solar radiation curve**)

Earthwatch Programme

A worldwide programme to monitor trends in the environment, established under the terms of the Declaration on the Human Environment in 1972. It is based on a series of monitoring systems and its activities are co-ordinated by UNEP. Activities also include research and assessments on the basis of data provided by the systems.

eccentricity

A measure of the deviation of an ellipse from a circle. It is the distance between the two foci divided by the length of the major axis. The eccentricity of the Earth's orbit at present is 0.016.

ecliptic

The great circle the Sun appears to describe on the celestial sphere. At present it is inclined at 23.5° to the Equator.

ecological balance

The components of a natural community are said to be in ecological balance if their relative numbers remain more or less constant, thus forming a stable ecosystem. Gradual readjustments to the composition of a balanced community continually take place in response to natural ecological successions and to other influences, including climate variations. People upset this balance by removing or introducing plants or animals, polluting the environment, by destroying habitats and by rapidly increasing the numbers of his own species.

ecological climatology

A branch of bioclimatology which studies the relationship between living organisms and their climatic environment. It includes the physiological adaptation of plants and animals to climate and the geographical distribution of plants and animals in relation to climate.

ecology
The scientific study of the inter-relationships among and between organisms, and between them and all aspects of their environment, living and non-living.

economic analysis of climate
Climate is an integral component of the natural resources of an area, and its change can thus affect many different economic activities. When treated as a resource, climate can reasonably be subjected to economic analyses applied to other natural resources. Many resource economists view climate as a public good, not subject to competitive market economics, and essentially free to all users. It is thus best analyzed as an exogenous variable whose fluctuations can affect resource flows, rates and costs of production, and the consumption of economic goods that are valued and distributed by market forces. In "centrally planned" and "managed" economies, climate remains an exogenous unknown for which allowances must be made in order to achieve desired production and consumption levels. In self-provisioning societies, climate can threaten the very survival of individuals and groups, and is best considered a risk or hazard.

economic summit meetings
Heads of government meetings which are attended by Canada, France, Japan, Italy, Russia, the United Kingdom and the United States. Also known as the Group of Seven. Other heads of governments may also attend by invitation.

At their Economic Summit meeting in Paris in July 1989, which represented a turning point in political attitudes towards environmental issues, the final Summit declaration strongly advocated "common efforts to limit emissions of carbon dioxide and other greenhouse gases". They went on to "strongly support the work undertaken by the IPCC", and voiced the "need to strengthen the worldwide network of observatories for greenhouse gases".

ecosphere
The layer of the Earth and the troposphere inhabited by, or suitable for, the existence of living organisms.

ecosystem
A term first used to describe the interdependence of species in the living world (the biome or community) with one another and with their non-living (abiotic) environment. Fundamental concepts include the flow of energy via food chains and food webs, and biogeochemical cycling of nutrients. Ecosystem principles can be applied at all scales – thus,

principles that apply to a pond, for example, apply equally to a lake, an ocean, or the whole planet.

ecosystem integrity
Ecosystem integrity refers to the state of health or wholesomeness of an ecosystem. It encompasses integrated, balanced and self-organizing interactions among its components, with no single component or group of components breaking the bounds of interdependency to dominate the whole.

ecotone
An ecosystem boundary, or transition zone between one ecosystem and another.

ECRC see **Environmental Change Research Centre**

edaphic
Of or pertaining to soil, especially with regard to its influence on plants and animals.

eddy
An individual transient element of turbulence, usually thought of as a translating and rotating air parcel.

eddy diffusion
The turbulent diffusion of properties in which eddies are considered to play a rôle analogous to that of molecules, but on a much larger scale.

Education and Training Programme of WMO
The Education and Training Programme of WMO promotes efforts in WMO Member countries to ensure that the necessary body of trained meteorologists, hydrologists, engineers and technicians is available. It is closely inter-related with all other major scientific and technical programmes of WMO.

EFTA
The European Free Trade Association.

EIS
The Energy and Industry Subgroup of IPCC WG III (Response Studies).

El Niño
A warm-water current which periodically flows southwards along the coast of Ecuador. It is associated with the Southern Oscillation (these effects are collectively known as an El Niño–Southern Oscillation, or ENSO, event) and has climatic effects throughout the Pacific region and

sometimes elsewhere. It occurs once every five to eight years, during the Christmas season (the name refers to the Christ child) when the prevailing trade winds weaken and the equatorial countercurrent strengthens. This causes warm surface waters, normally driven westwards by the wind to form a deep layer off Indonesia to flow eastwards to overlie the cold waters of the Peru current. In exceptional years (e.g. 1891, 1925, 1953, 1972–3 and 1982–3) the extent to which the upwelling of the nutrient-rich cold waters is inhibited causes the death of a large proportion of the plankton population and a decline in the numbers of surface fish. The El Niño may also cause heavy rainfall along the otherwise dry coasts of Ecuador and Peru. The "negative" phase of an El Niño event is called a La Niña.

EMEP
An ECE co-operative programme which was formed in 1977 for the monitoring and evaluation of the long-range transmission of air pollution in Europe.

emission control measures
Those human activities that assist in controlling the net addition of greenhouse gases to the atmosphere, which include measures to develop new sinks for such gases beyond those that occur naturally without human interference.

emission factor
The relationship between the amount of pollution produced and the amount of raw material processed.

emission levels per capita
The emission levels per person in a country.

emission levels per energy consumption per consumption unit
The emission levels in a country expressed as a percentage of the energy consumption per unit of production in that country.

emission levels per size of the natural carbon sinks
The emission levels in a country expressed as a percentage of the size of the carbon sink in that country.

emission levels per unit of GDP
The emission levels in a country expressed per a common financial unit of the gross domestic product.

emission levels per unit of GNP
The emission levels in a country expressed per a common financial unit

of the gross national product.

emission rate
The amount of pollutant emitted per unit of time.

emission standard
The maximum amount of pollutant permitted to be discharged from a single polluting source.

emission scenarios (IPCC)
The Response Strategies Working Group of the IPCC developed emissions scenarios for evaluation by the IPCC Working Group 1. The scenarios were developed by an expert group using the Atmospheric Stabilization Framework of the US Environmental Protection Agency (EPA) and the Netherlands National Institute for Public Health and Environmental Protections (RIVM). The scenarios cover the emissions of carbon dioxide, methane, nitrous oxide, chlorofluorocarbons, carbon monoxide and nitrogen oxides from the present up to the year 2100. Growth of the economy and population was taken to be common for all scenarios. Population was assumed to approach 10.5 billion in the second half of the next century. Economic growth was assumed to be 2–3% annually in the coming decade in the OECD countries and 3–5% in the Eastern European and developing countries. The economic growth levels were assumed to decrease thereafter. Four scenarios were considered:

Scenario A (Business as Usual):
In scenario A, the energy supply is coal intensive, and on the demand side only modest efficiency increases are achieved. Nevertheless, the integration of base case studies by the Working Group 3 Energy and Industry Subgroup has higher energy consumption and associated emissions. Carbon monoxide controls are modest, deforestation continues until the tropical forests are depleted and agricultural emissions of methane and nitrous oxide are uncontrolled. For CFCs the Montreal Protocol is implemented, albeit with only partial participation.

Scenario B:
In scenario B, the energy-supply mix shifts towards lower carbon fuels, notably natural gas. Large efficiency increases are achieved. Carbon monoxide controls are stringent, deforestation is reversed and the Montreal Protocol implemented with full participation.

Scenario C:
In scenario C, a shift towards renewables and nuclear energy takes place in the second half of the 21st century. CFCs are now phased out and agricultural emissions limited.

Scenario D:
In scenario D, a shift to renewables and nuclear energy in the first half of the 21st century reduces the emissions of carbon dioxide,

initially more or less stabilizing emissions in the industrialized countries. The scenario shows that stringent controls in industrialized countries combined with moderated growth of emissions in developing countries could stabilize atmospheric concentrations.

emissions
Materials such as gases, particles, vapours and chemical compounds that come out of smokestacks, chimneys and tailpipes.

endemic
Refers to plant or animal species confined to a well defined area.

energy efficiency
A measure of the energy required to perform a specific task or produce a specific product. It is a useful parameter for comparing the potential ability to do the same thing with less energy. Comparisons between nations of the potential for improvement in energy use should more appropriately focus on energy efficiency, and not energy intensity.

energy intensity
The aggregate energy used per unit of activity such as energy use per GDP. For example, an energy-intensive activity, such as steel production, requires high use of energy relative to its labour force, investment and level of output, while an activity with low energy intensity (e.g. most consumer service activities) has comparatively low energy use relative to these variables.

energy of storms
In a tropical or subtropical cyclone, the energy released as latent heat from the rain is equivalent to between 5 and 100 hydrogen bombs a day. This causes the conversion of gravitational potential energy, equal to between 1 and 10 hydrogen bombs a day due to the sinking of cold air under warm air, into wind systems.

energy potentials
Four different types of energy potential can be distinguished:
* Theoretical potential is the sum of energy sources and energy conservation, calculated on the basis of physical or scientific laws.
* Technical potential is derived from the theoretical potential, taking into consideration the efficiency of the various systems used to tap energy sources, or to conserve energy, as well as other relevant technical aspects. Technical potential represents state-of-the-art technology.
* Economic potential restricts the technical potential to that part which compares the costs of competing systems in terms of their relative economic competitiveness.

* Expected potential is that portion of the economic potential which takes into consideration the speed of market introduction and other factors, and hence is the utilized economic potential to be expected within a given general economic setting.

engineering hydrology
The branch of applied hydrology which deals with hydrological information intended for engineering applications such as planning, designing, and operating and maintaining engineering measures and structures.

enhanced greenhouse effect
The increased intensity of the natural greenhouse effect attributable to increases in atmospheric greenhouse gas concentrations induced by human activity. It is important to distinguish clearly between the natural greenhouse effect, which is essential for life on Earth, and the effect of humans on that natural effect through increased atmospheric concentrations of greenhouse gases.

ENSO events see El Niño

environment
The sum of all external conditions affecting the life, development and survival of an organism.

Environmental Change Research Centre (ECRC)
Based at University College London, the Centre is concerned with reconstructing, monitoring and modelling trends in aquatic and terrestrial ecosystems in relation to problems of acid deposition, surface water acidification, eutrophication, stream- and river-water quality, climate change and nature conservation. Current contracts include sulphur deposition and critical loads for UK freshwaters, acidification of alpine lakes, climate change inferences from salt-lake sediments in North America, eutrophication of Antarctic lakes, upland peat erosion in the UK, and recent environmental change in Lake Baikal.

Environmental Change Unit
An interdisciplinary unit at the University of Oxford (UK) dedicated to collaborative research on the nature, causes and impact of environmental change, and to the development of management strategies to deal with future changes.

environmental impact assessments
An interdisciplinary process by which the environmental consequences of proposed actions and various alternatives are presented and considered as an aid to decision-making. The process may involve

assessments of ecological, sociological, anthropological, economic, geological and other environmental impacts. In many countries, industrial organizations planning new projects are required by law to conduct such studies and to produce an environmental impact assessment which can then be examined critically.

environmental refugees

People forced to leave their homeland and because of environmental factors such as a rising sea level or a significant change in the climate.

EOS

The Earth Observing System.

ephemeral lake

A lake which becomes dry during the dry season or in particularly dry years.

equatorial easterlies

Easterly winds of the equatorial zone, of great vertical extent, in the regions of the junction of the trade winds of the Northern and Southern Hemispheres.

equilibrium and realized climate change (IPCC WG I: Policymakers Summary)

When the radiative forcing on the Earth–atmosphere system is changed, for example by increasing greenhouse gas concentrations, the atmosphere will try to respond (by warming) immediately. But the atmosphere is closely coupled to the oceans, so in order for the air to be warmed by the greenhouse effect, the oceans also have to be warmed; because of their thermal capacity, this takes decades or centuries. This exchange of heat between atmosphere and ocean will act to slow down the temperature rise forced by the greenhouse effect.

In a hypothetical example where the concentration of greenhouse gases in the atmosphere, following a period of constancy, rises suddenly to a new level and remains there, the radiative forcing would also rise rapidly to a new level. This increased radiative forcing would cause the atmosphere and oceans to warm, and eventually come to a new, stable temperature. A commitment to this equilibrium temperature rise is incurred as soon as the greenhouse gas concentration changes. But at any time before equilibrium is reached, the actual temperature will have risen by only part of the equilibrium temperature change, known as the realized temperature change.

equitability

The assessment of the fairness of a measure in its distribution of favourable or unfavourable impacts across the economic, environmental, social and political interests that are affected.

ERBE
The Earth Radiation Budget Experiment.

ERS
An earth resources satellite.

EUMETSAT
The European organization which operates METEOSAT. EUMETSAT stands for the European Organization for the Exploitation of Meteorological Satellites.

euphotic zone
The layer of a body of water that receives sufficient sunlight for photosynthesis. The depth of this layer, which is about 80m, is determined by the water's extinction coefficient, the cloudiness and the sunlight's angle of incidence.

eutrophic lake
A lake characterized by a great amount of nutrient and biogenic matter and by highly developed phytoplankton in summer.

eutrophication
The slow ageing process of a lake evolving into a marsh and eventually disappearing. During eutrophication the lake is choked by abundant plant life. Human activities that add nutrients to a water body can increase the rate of eutrophication.

evaporation
The process by which a liquid is changed into a gas. In the atmosphere it is the physical process by which water is transferred from the Earth's surface to the atmosphere through the evaporation of water or ice into water vapour and through transpiration from plants. Evaporation is a very important aspect of the Earth's energy balance; by this process, surplus energy in tropical areas is transferred by the flux of water vapour to the radiation-deficit areas towards polar areas.

evaporation pan
An open, cylindrical metal pan mounted on a flat wooden base. Daily readings of the water level in the pan and the amount of precipitation from a nearby rain-gauge enable the amount of evaporation to be calculated.

evaporimeter
An instrument for measuring the amount of water evaporated into the

atmosphere during a given time interval.

evapotranspiration (actual)
The total quantity of water vapour evaporated by a surface and transpired by vegetation, when the surface and vegetation is at its actual moisture content (see also **potential evapotranspiration**).

experimental basin
An hydrological basin in which natural conditions are deliberately modified and in which the effects of these modifications on the hydrological cycle are studied.

exploitation of climate
The use of a climate resource in order to maximize its benefits and minimize its costs or adverse effects. All countries endeavour to "exploit" their national climate resources, but clearly some countries are in a better position to do so than others.

extraterrestrial regions
This term usually refers to the non-land areas and in particular the oceans, but may also refer to "areas" outside the Earth.

extreme temperatures
Highest and lowest temperatures attained during a given time interval.

extreme-value distribution
A parameter used in statistics to indicate the probability distribution of the largest and/or smallest observation in a sample. Commonly used extreme-value distributions include those of Frechet, Gumbel and Weibull.

F

FAGS see **Federation of Astronomical and Geophysical Data Analysis Services**

FAO see **Food and Agriculture Organization**

fallout
A term which may be applied to both the process of deposition of solid material on the Earth's surface and to the deposited material itself. It may be used in such a sense as to signify only "dry deposition" (mainly the result of gravitational settling): in such a sense it is used in contrast

to the term "wet deposition" caused by precipitation, which is sometimes call "washout". The term fallout has, however, been used mainly in respect of the radioactive debris associated with a nuclear explosion.

FCCC
The Framework Convention on Climate Change.

Federation of Astronomical and Geophysical Data Analysis Services (FAGS)

FAGS was formed by ICSU in 1954 and it includes ten Permanent Services, each operating under the authority of one or more of the interested Unions: IAU, IUGG and URSI. Their tasks are to continuously collect observations, information and data related to astronomy, geodesy, geophysics and allied sciences; to analyze, synthesize and draw conclusions from them; to distribute data; and to publish the results obtained.
The Unions responsible for the Permanent Services are:
* International Earth Rotation Service (IAU–IUGG–URSI)
* International Service of Geomagnetic Indices (IUGG)
* Bureau Gravimétrique International (IUGG)
* Quarterly Bulletin on Solar Activity (IAU)
* Permanent Service for Mean Sea Level (IUGG)
* International Centre for Earth Tides (IUGG)
* International Ursigram and World Days Service (URSI-IAU-IUGG)
* World Glacier Monitoring Service
* Centre de Données Stellaires (IAU-IUGG)
* Sunspot Index Data Centre (SIDC).

feedback mechanisms

Mechanisms or processes that result in the response to an initial impulse or action being different than it would be directly. These mechanisms may amplify the final impact or the initial action (known as positive feedback), or may reduce the final impact or the initial action in which case it is known as negative feedback.

feedbacks (atmospheric)

It is relatively easy to predict the direct effect of the increased radiative forcing due to increases in greenhouse gases. However, as climate begins to warm, various processes act to amplify the warming (through positive feedbacks) or reduce it (through negative feedbacks). The main feedbacks which have been identified are due to changes in water vapour, sea ice, clouds and the oceans. The best tools available which take all the relevant processes into account are three-dimensional mathematical models of the climate system (atmosphere–ocean–ice–land), known as general circulation models (GCMs). These models synthesize our knowledge of the physical and dynamical processes in the overall system and allow for the complex interactions between the various components. However, in their current state of development the models are comparatively crude in their description of many of the processes involved.

FID see **International Federation for Information and Documentation**

field capacity

The mass of water retained by a previously saturated soil when free drainage has ceased, known as the soil's field capacity or water-holding capacity.

fingerprint method

A key issue in climate detection studies is the need to be able to attribute the observed changes in climate (or part of them) to the enhanced greenhouse effect. Confidence in the attribution is increased as predictions of changes in various components of the climate system are borne out by the observed data in more and more detail. The method proposed for this purpose is the "fingerprint method"; namely, identification of an observed multivariate signal with a structure unique to the predicted enhanced greenhouse effect, where a multivariate signal changes in a single climate element (such as temperature) at many places or levels in the atmosphere, or changes in a number of different elements, or changes in different elements at different places.

first detection

The identification of a "precursor signal", detectable above the "noise" of natural climatic variability, of a significant change in a climate parameter, and attribution of this change to an increase in atmospheric carbon dioxide concentration, or some other human-induced change. The signal may be estimated by numerical modelling of the climate, and the noise can be estimated using instrumental data. For any modelled signal that is estimated, the corresponding noise can be estimated from observational data, and a signal-to-noise ratio can be calculated to provide a quantitative measure of detectability.

First International Polar Year (1882–83)

The First Polar Year in 1882–83 was a joint effort of twelve countries (Austria/Hungary, Denmark, Finland, France, Germany, Netherlands, Norway, Russia, Sweden, United Kingdom, Canada and USA) to establish and operate 14 stations surrounding the North Pole. Observations were made in connection with meteorology, geomagnetism, auroral phenomena, ocean currents and tides, structure and motion of ice, atmospheric electricity and air samples for analysis, etc. Over 40 observatories were involved in the effort by undertaking expanded programmes of observations.

First World Climate Conference (1979)

In 1974, the WMO Executive Council agreed that WMO should initiate an international programme on climate. A panel of experts was created, and it quickly co-opted representatives of other interested organizations to lay the foundations of a new climate programme. The proposals made by the panel were discussed at various forums and it was decided that WMO should establish a wide-ranging World Climate Programme. A major milestone in the development of the new programme was the First World Climate Conference, convened by WMO in February 1979. The problem of possible human influence on climate was recognized as an issue of special importance.

In an appeal to nations, the Conference Declaration stressed the urgent need for the nations of the world:
* to take full advantage of man's present knowledge of climate;
* to take steps significantly to improve that knowledge; and
* to foresee and to prevent potential manmade changes in climate that might be adverse to the wellbeing of humanity.

The Eighth World Meteorological Congress, meeting in 1979 shortly after the First World Climate Conference, formally created the World Climate Programme.

flash floods

A very rapid rise of water with little or no advance warning, most often when an intense thunderstorm causes very high rainfalls on a fairly small area in a very short space of time.

flood proofing

Techniques for preventing flood damage in a flood-hazard area.

flue

Any passage designed to carry combustion gases and entrained particulates.

flue gas

The air and pollutants emitted to the atmosphere after a production process or combustion takes place.

flue-gas scrubber

A type of equipment that removes fly-ash or other objectionable materials from flue gas by using sprays, wet baffles, or other means that require water as the primary separation mechanism.

fluorocarbons

A gas used as a propellant in aerosols.

fog
The suspension of very small water droplets in the air, generally reducing the horizontal visibility at the Earth's surface to less than 1km.

föhn wind
A warm, dry wind on the lee side of a mountain range, the warmth and dryness of the air being due to the adiabatic compression air upon descending the mountain slopes.

Food and Agriculture Organization (FAO)
FAO was one of the first UN specialized agencies to be formed, and is one of the largest. Based in Rome, its main aim is to increase food production and availability throughout the world. The Conference of FAO, which meets every two years, determines the policy and approves the budget. The Council of FAO meets at least three times between thte regular sessions of the Conference, and has several Standing Committees.

Food and Agriculture Organization: Climate and Climate Change Programmes
FAO has a long history of involvement with climate-related problems. Through a first formal agreement with WMO, the Inter-agency Agro-climatology Project was created in 1960, to promote agro-climatological studies in areas where large agricultural developments were anticipated. Several detailed regional studies on climate/agriculture interactions were subsequently published, starting in 1962. In 1968, FAO, WMO and UNESCO established the Inter-agency Group on Agricultural Biometeo-rology, which was to become part of the regular programme of FAO as the Agrometeorology Group or Agroclimate Unit. The Stockholm Conference on the Environment in 1972, and the 1974 World Food Conference constituted turning points in the perception of crop–climate relations in FAO. They led to recognizing the prevalent effect of climate on the variability of food supply, and the Agroecological Zones Project (AEZ) was set up to estimate the food production potential of developing countries. The results of this major undertaking, co-ordinated by the Land and Water Development Division of FAO, include not only regional publications but also the elaboration of a general methodology and the accumulation of a valuable set of reference climatic datasets published between 1984 and 1987.

The FAO Working Group on Climate Change whose 13 members belong to the FAO Departments of Agriculture, Fisheries, Forestry, Economic and Social Policy, and the Legal Office, constitutes the first direct involvement of FAO with climate change per se: it is a technical group established in February 1988 to review the available evidence and provide an assessment of the possible impacts on world agriculture and food production. In 1990, reflecting the emphasis FAO is putting on

sustainable development, the Working Group was formally absorbed into the Inter-Departmental Working Group on Environment and Sustainable Development as one of several ad hoc working groups, with the following terms of reference:
* as a co-ordination mechanism for climate change-related work with FAO;
* to provide a forum for assessing internal and external work on climate change and its implication for FAO's Regular and Field Programmes;
* to formulate recommendations for FAO's Programme of Work and Budget;
* to assist in the clearance of IPCC and other UN documents related to climate change and its potential impact on agriculture, forestry and fisheries;
* to revise and elaborate the FAO Position Paper on Climate Change.

The following lists the technical studies currently undertaken by FAO as regular programme activities:
* guidelines and methods for assessment of climate change impact on irrigated agriculture;
* methodology development for regional impact studies on crop agriculture;
* map and classification of the world's low-lying coastal areas;
* identification of sensitive inshore areas and wetlands;
* monitoring of fisheries ecosystems;
* the global forest resources assessment;
* future development of tropical forest resources;
* fire statistics from developing countries;
* worldwide status of forest decline;
* joint FAO–IAEA study on biogeochemical cycles;
* AMDASS: the Agrometeorological Data Systems.

forecast lead time
The interval of time between the issuing of a forecast and the expected occurrence of the parameter that is forecast.

forecasting by analogy
To understand how societies might best respond to a yet unknown change in regional as well as global climate regimes, it would be highly desirable to know how societies have been affected by and how they coped with the effects of extreme meteorological events, such as droughts, that have occurred in the recent past. Although the climate in the future might not be like that of the recent past, one can assume that, barring unforeseeable shocks to social systems, societal institutions in the near future will be like or nearly like those of the recent past. By identifying societal strengths and weaknesses in past responses to extreme meteorological events, societies can act in a much more informed manner to eradicate the weaknesses and capitalize on existing strengths. They can then better prepare for the implications of an

uncertain climate future.

Forecasting societal responses by analogy can be viewed as a "win/win" situation. Focusing on coping with extreme meteorological events produces improved responses to such events, whether or not the climate of the future is different from that of the recent past. Such research therefore results in an improved understanding of the interactions between climate variability and society.

forest management see **afforestation, deforestation and reforestation**

fossil fuel
Any hydrocarbon deposit that can be burned for heat or power, such as petroleum, coal and natural gas.

fossil water
Water infiltrated into an aquifer during an ancient geological period under climatic and morphological conditions different from the present, and stored since that time.

Framework Convention on Climate Change see **Intergovernmental Negotiating Committee on a Framework Convention on Climate Change** and **Climate Change Convention**

freezing rain
Supercooled water drops of drizzle, or rain, which freeze on impact to form a coating of ice upon the ground and on the objects they strike.

front
The interface or boundary between two different air masses which have originated from widely separated regions. A cold front is the leading edge of an advancing cold air mass, while a warm front is the trailing edge of a retreating cold air mass.

frost-free season
The period between the last frost in the spring and the first frost in the autumn (fall).

G

GACC see **General Agreement on Climate Change**

GADS see **Global Aerosol Data System**

Gaia hypothesis
The hypothesis that the Earth's physical and biological systems are considered to be a complex and self-equilibriating entity.

GATT see **General Agreement on Tariffs and Trade**

GAW see **Global Atmosphere Watch**

GCIP
The GEWEX Continental-Scale International Project.

GCM see **general circulation model**

GCOS see **Global Climate Observing System**

GCTE see **Global Change and Terrestrial Ecosystems Project**

GDP see **gross domestic product**

GDPS see **Global Data Processing System**

GEDEX
The Greenhouse Effect Detection Experiment.

GEMS see **Global Environment Monitoring System**

General Agreement on Climate Change (GACC)
A proposed/possible agreement on climate change which would consist of core agreements on allowable national contributions to global warming over time. It would facilitate a wide range of other agreements on technology transfer, funding mechanisms and other issues as needed to accommodate the interests of all nations.

General Agreement on Tariffs and Trade (GATT)
A multinational treaty concluded within the scope of the United Nations with the goal of eliminating obstacles to international trade. GATT entered into force on 1 January 1948. GATT includes the following fundamental principles: (a) trade between countries must take place on the basis of non-discrimination; (b) all signatory states are obliged to observe the most favoured nation principle when levying import and export duties and taxes; (c) domestic industries are to be protected exclusively by tariffs; (d) quotas and other non-tariff trade barriers are generally inadmissible and are permitted only in certain exceptional cases, such as to protect a country's balance of payments.

general circulation model: Canadian
One of the most advanced GCMs in use today is that developed at the

Canadian Climate Centre in Toronto. The model incorporates improvements over GCMs used in earlier climate modelling experiments by other groups. In particular, it provides a much higher spatial resolution (that is a finer grid) than previous GCMs, giving more than twice the information coverage and allowing a much more detailed representation of local climatic processes. It also provides a more accurate simulation of the reflective and absorptive properties of clouds, annual and daily solar heating, ocean temperature and ice boundaries. In common with most other models, however, it lacks a fully interactive, circulating ocean. Work is continuing on additional improvements, including a two-dimensional dynamic ocean.

general circulation of the atmosphere

The average, worldwide system of winds and associated weather systems. Air movement is caused by differential heating of the Earth's surface and atmosphere and by the Earth's rotation, with topographic differences causing local variations.

geographic information system (GIS)

A computer-based "tool" which captures, displays and manipulates geographically referenced data.

geological era

The primary and largest division of geological time. Limits are rather arbitrary, but each begins and ends with a time of major crustal, climatic and volcanic upheaval in some part of the Earth, with a large worldwide withdrawal of the sea from land masses. Five geological eras are recognized: Archeozoic, Proterozoic, Palaeozoic, Mesozoic and Cenozoic. Some authorities regard the Cenozoic as two eras, the Tertiary and Quaternary. All eras are divided into at least two geological periods and a number of geologic epochs.

Geophysical Fluid Dynamics Laboratory (GFDL)

Located at Princeton University, USA.

geostrophic wind

A horizontal wind whose direction and speed are determined by a balance between the force due to the Earth's rotation (the Coriolis effect) and the pressure-gradient force.

geothermal energy

Geothermal energy is energy, as heat, derived from anomalies in the temperature gradient in the Earth's crust. Normally, temperature in the crust increases with depth at a constant rate, but locally water or rock may be much hotter than the surrounding rocks. The hot water may then be tapped and its heat used. Geothermal energy is delivered as hot water and can be used only close to its source. It is not inexhaustible, since the extraction of heat from the localized anomaly causes cooling, so that eventually the source is cooled to the temperature of the surrounding material.

GESAMP see **Joint Group of Experts on the Scientific Aspects of Marine Pollution**

GEWEX see **Global Energy and Water Cycle Experiment**

GFDL see **Geophysical Fluid Dynamics Laboratory**

GIEWS see **Global Information and Early Warning System on Food and Agriculture**

giga-
A prefix meaning billion (1×10^9).

GIS see **geographic information system**

GISS see **Goddard Institute for Space Studies**

glacial maximum
The greatest extent of Pleistocene ice which existed about 200,000 years ago, during the penultimate glacial period. At that time ice extended across the Arctic Ocean and joined with the ice sheets that covered much of North America, Greenland, Northern Russia and North West Europe. The ice is believed to have extended across the North Sea to engulf all but the very south of the British Isles. At the glacial maximum an area of some 46 million km^2 was beneath ice, over three times that of any "glacial" period.

glacial period
Generally, an interval of geologic time which was marked by a major equatorward advance of ice. The term may be applied to an entire ice age or to the individual glacial "stages" which make up an ice age.

glacier
A large mass of ice, resting on or adjacent to a land surface. Glaciers may be classified in several ways. The most useful division is based on temperature, and three categories are recognized. In temperate (or warm) glaciers (e.g. those of the Alps) the ice is at pressure melting point throughout, except during winter when the top few metres may be well below 0°C. Movement is largely by basal slip. Polar (or cold) glaciers (e.g. parts of the Antarctic sheet) have temperatures well below the pressure melting point, and movement, which is slow, is largely by internal deformation. Subpolar glaciers (e.g. those of Spitzbergen) have

temperate interiors and cold margins and so are composite in nature.

glacier flood
A sudden outburst of water released by a glacier.

glacier movements – the past 100 years
Most valley glaciers have been retreating over the past hundred years. Although long records of glacier length are available mainly for some glaciers in the European and North Atlantic region, geomorphological investigations have shown that the trend of glacier retreat has generally been worldwide since the Little Ice Age. Wastage was most pronounced in the middle of the 20th century. Around 1960, many glaciers started to advance. In the 1980s this advance slowed down or stopped in several glacier basins, but not everywhere.

glaciology
The scientific study of ice in all its forms. It therefore includes the study of ice in the atmosphere, in lakes, rivers and oceans, on and beneath the ground. Commonly, however, it is the study of glaciers.

Global Aerosol Data System (GADS)
An Experts Meeting on a proposed Global Aerosol Data System (GADS) was held in 1990. The purpose of the meeting was to bring together experts both in atmospheric aerosol measurements and impacts, and in database management systems, to assess the current state of the art, and to make recommendations for developing a co-ordinated plan for future work in the aerosol data management aspects of the International Global Aerosol Programme (IGAP). A plan outline for IGAP has been prepared and is in the process of being further developed by the various sub-groups of the Joint Working Group on International Aerosol Climatology Project (IACP), which is sponsored by the three International Association of Meteorology and Atmospheric Physics (IAMAP) Commissions: Clouds and Precipitation (ICCP), Radiation (IRC), and Atmospheric Chemistry and Global Pollution (ICACGP).

Our ability to assess the impacts of atmospheric aerosols on the quality of life (e.g. health, environment and climate) is constrained by our limited knowledge of their physicochemical, spatial–temporal, and source–sink cycle transformation patterns. This deficiency is partly due to the fact that we do not have convenient access to existing data on atmospheric aerosols that reside at different locations throughout the world. The idea of establishing a computer-accessible global aerosol database has been discussed for over ten years and the rationale for the establishment of GADS can be formulated from three points of view:
* *Scientific:* Aerosols play an important rôle in climate and biogeochemical cycling of trace substances that affect the environment and health. Currently, major global-scale aerosol databases exist at different institutions around the world. However, these need to be

integrated, reconciled, packaged and made available to potential users.

* *Geopolitical:* The geopolitical climate is conducive to developing co-ordinated research efforts in the area of global change. The Global Atmospheric Watch of WMO has as one of its components the task to arrange for systematic aerosol measurements and provide data and analysis. It is likely that during the 1990s, major international research programmes and environmental protection programmes will be initiated that could beneficially utilize GADS.
* *Technological:* The trends in computer, communication and software technologies that occurred in the 1970s and 1980s make a global aerosol data-exchange system feasible in the 1990s.

The conceptual design approach of the Global Aerosol Data System is that it should evolve through three phases:

* the creation of an Aerosol Master Directory;
* demonstration of Data Access and Delivery Systems (DADS);
* implementation of an operational Global Aerosol Data System (GADS).

Global Atmosphere Watch (GAW)

Political events in 1989 clearly indicated that governments had become increasingly concerned about the atmosphere and the possible long-term changes in its composition arising from human activities. At its 41st session in 1989, WMO's Executive Council approved measures for strengthening WMO's rôle in the scientific aspects of this global problem. The WMO has much experience in related activities, having set up and run for many years systems for monitoring the composition of the global atmosphere such as BAPMoN and GO_3OS. Continued involvement in the planning and organizing of such systems and in the preparation of related scientific assessments is a major responsibility of WMO and it was for this reason that the concept of the Global Atmosphere Watch was proposed and developed.

Much as the World Weather Watch (WWW) covers the provision of weather and climate information, GAW will provide atmospheric composition information. WMO Members will be encouraged to make atmospheric chemical observations on a continuous basis as an integral part of their atmospheric observation programmes, and with the same care and operational regularity as for meteorological observations, in the course of the 1990s.

Some WMO Members already provide vital centralized data-collection points and services for GAW: e.g. Canada operates the WMO World Ozone Data Center in Toronto and publishes ozone data every other month, and the USA hosts a centre for data on precipitation chemistry, acid rain and atmospheric turbidity (transparency) measurements. In September 1989, agreement was reached with the Japan Meteorological Agency in Tokyo that it should be designated the WMO World Data Centre for Greenhouse Gases.

Considerable efforts are being made to inform WMO Members and the general public about GAW. Press releases and fact sheets have been issued and others (on precipitation chemistry, carbon dioxide and long-range transport of pollutants) are being prepared.

The anticipated development of GAW during the period 1992–2001 is aimed at meeting present and future data requirements of WMO Members and international organizations for protecting and managing the global atmospheric environment, was taken into account when drafting the WMO Third Long-term Plan.

The WMO–GAW System can be seen as the umbrella organization for the Background Air Pollution Monitoring Network (BAPMoN) and the Global Ozone Observing System (GO₃OS). The GAW addresses both monitoring and research activities involving atmospheric composition-related measurements. Results are used, among other things, to serve as an early-warning system to detect changes in the transport of pollutants, changes in the concentrations of greenhouse gases and changes in the ozone layer.

The present GAW network consists of 196 monitoring stations. Of these many have varying functions, including:
* 152 which measure precipitation chemistry;
* 90 BAPMoN stations which measure atmospheric turbidity;
* 84 which measure suspended particulate matter;
* 23 which measure carbon dioxide concentrations;
* 22 which measure surface ozone;
* 7 which monitor methane; and
* 5 which monitor CFCs.

global baseline datasets

Global baseline datasets are designed to provide a complete historical record of the climate. Such datasets should include, among other things, data on the following parameters: land and ocean surface temperatures, high-latitude winter temperatures, upper tropospheric temperatures, stratospheric temperatures, sea-level height, tropospheric water-vapour content, summer continental moisture, cryospheric elements (snow, sea ice, freeze-up, break-up), circulation parameters, outgoing longwave radiation, atmospheric pressure, surface- and upper-air relative humidity and proxy data.

A meeting of experts convened by WMO was held in November 1990 to make recommendations on preparing a global dataset for climate change studies. During 1990 work began on the construction of a comprehensive upper-air dataset and a dataset of so-called metadata (e.g. historical information about observing stations and instrumentation).

Global Change and Terrestrial Ecosystems Project (GCTE)

The Global Change and Terrestrial Ecosystems Project is a core project of the IGBP, aimed at developing the capacity to predict the effects of changes in climate, atmospheric composition and land-use practices on terrestrial ecosystems. This capacity is required both because the ecosystem changes are of direct importance to humans, and because they will have a feedback effect on evapotranspiration, albedo and surface roughness. The project has two main foci: ecosystem "physiology" – the exchanges of energy and materials, and their distribution and storage; and ecosystem structure – the changes in species (functional type) composition and physiognomic structure, on the patch, landscape and regional (continental) scales. The project is based on close integration of experimentation and modelling. It consists of seven core activities, each of which is made up of a number of particular tasks, which include such topics as elevated CO_2 effects on ecosystem functioning, changes in biogeochemical cycling, and soil and vegetation dynamics.

Global Climate Observing System (GCOS)

An observing system/programme recommended by the Scientific/ Technical part of the Second World Climate Conference, and endorsed by the Eleventh Congress of WMO which met in Geneva in May 1991. The WMO–IOC–ICSU sponsored Global Climate Observing System (GCOS) is intended to meet the needs for:
* climate system monitoring, climate change detection and monitoring of the response to climate change, especially in terrestrial ecosystems and mean sea level;
* data for application to national economic development;
* research towards improved understanding, modelling and prediction of the climate system.
GCOS will build, as far as possible, on existing operational and scientific observing, data management and information distribution systems, and further enhancement of these systems. GCOS will be based upon:
* improved World Weather Watch systems and the Integrated Global Ocean Services System;
* data communication and other infrastructures necessary to support operational climate forecasting;
* the establishment of a Global Ocean Observing System (GOOS) for physical, chemical and biological measurements;
* the maintenance and enhancement of programmes monitoring other key components of the climate system, such as the distribution of important atmospheric constituents (including the Global Atmosphere Watch), terrestrial ecosystems (including the International Geo-sphere–Biosphere Programmes), as well as clouds and the hydro-logical cycle, the Earth's radiation budget, ice sheets and precipitation over the oceans (including the World Climate Research Programme).
A Memorandum of Understanding between WMO, IOC and ICSU was signed on 24 October 1991, and came into force on 1 January 1992. WMO, IOC and ICSU, as the initial sponsoring organizations agree:
* to co-operate in organizing a Global Climate Observing System (GCOS), based on the co-ordination of existing or planned operational

and research programmes for observing the global climate system, and the further development of these programmes as required to ensure continuity of observations;
*	that the GCOS shall have, as its long-range objectives, to support all aspects of the World Climate Programme and relevant aspects of other climate-related global programmes, specifically to meet the data needs for climate system monitoring and applications to national economic development, as well as research leading to improved understanding, modelling and prediction of the climate system;
*	to consult and call upon other relevant national and international agencies, institutions and organizations, to collaborate in the organization and participate in the implementation of the GCOS;
*	to establish a Joint Scientific and Technical Committee (JSTC), to provide scientific and technical guidance for the organization and further development of the GCOS, and a Joint Planning Office;
*	to support, through administrative and financial arrangements, the activities of the JSTC for GCOS and its planning staff.

global climate system

Climate is ultimately the product of a complex web of physical, chemical and biological processes that take place within the Earth–atmosphere system in response to the radiative energy input from the Sun. The essential components of the global climate system are:
*	The atmosphere, which is the most rapidly varying part of the system. The troposphere has a characteristic response or thermal adjustment time of the order of a week, while the stratosphere and higher layers of the atmosphere have quite different processes and timescales.
*	The ocean, which interacts with the overlying atmosphere or ice on timescales of months to years, while the deeper ocean has a thermal adjustment time of the order of decades to centuries.
*	The land surface, which comprises the land masses of the continents, including the lakes, rivers and groundwater, which are important components of the hydrological cycle. These are variable components of the climate system on all timescales.
*	The cryosphere, which comprises the continental ice sheets, mountain glaciers, sea ice, surface snow cover and permafrost zones. The changes of snow cover and the extent of sea ice show large seasonal variations while the glaciers and ice sheets respond much more slowly.
*	The biosphere, which is the collective term for all living and dead organic matter in our environment. The fraction of the biosphere which is most significant for shaping climate on seasonal to decadal timescales is the terrestrial vegetation, while phytoplankton in the upper ocean is a deciding factor on longer timescales.

Global Data Processing System (GDPS)

The Global Data Processing System is a key part of the World Weather Watch (WWW). Data-processing requirements for meteorological services vary considerably, particularly among climate regions or subregions. The requirements for data and products, based mainly on experience and studies in the more industrialized countries, include end-user requirements for medium- and long-range forecasts (e.g. planning outlooks). Several user activities are extremely weather- sensitive in their longer-term planning, and the requirements for weather outlooks for one week and for one season ahead are growing rapidly. Medium- and long-range forecasts will become increasingly significant in the 1990s and beyond.

Climate monitoring requirements for data and products include special computations and analyses done operationally as part of the GDPS, but with special procedures specified through the World Climate Programme.

The World Weather Watch Data Management (WWWDM) function will be important for the further integration of the GDPS functions and activities into the WWW system. It will ensure production of high-quality real-time observational data in the GOS and maintenance of this quality during data transport on the GTS. It will also provide WMO Members with the ability to access both WWW observational data and processed products, on a continuous and reliable basis.

The Global Data Processing System provides a systematic day-to-day capability to synthesize real-time data from a variety of sources into global analysis fields that can satisfy many climate monitoring requirements as well as weather-forecasting requirements. Experts from both the GDPS and climate communities will be required to validate requirements, consider possible ways for the GDPS to satisfy the requirements, and develop the necessary reports and proposals to seek commitments from WMO Members and appropriate GDPS centres (WMCs or RSMCs) to undertake analyses that contribute directly to climate monitoring. Mechanisms to exchange climate monitoring products over the GTS will also be developed.

GDPS centres are expected to become more involved in climate data collection and provide input to real-time climate diagnosis (e.g. 10-day and 30-day summaries, and monitoring of the El Niño and drought situations). This activity will be achievable by automated GDPS centres which have databases available for hemispheric or global models. There should also be continuing discussions among the WMO Technical Commissions (CBS, CAeM, CCl and CAgM) to formulate requirements for specialized WWW services to other WMO programmes. After the technical commissions have determined the requirements, including the WWW facilities and services needed, the necessary procedures could be included in the operational WWW system at national or regional levels.

Global Energy and Water Cycle Experiment (GEWEX)

The Global Energy and Water Cycle Experiment (GEWEX) is a programme launched by the WCRP to observe, understand, and model the hydrological cycle and energy fluxes in the fast component of the climate system. GEWEX has been formally endorsed by WMO and ICSU as a major component of the international climate research strategy. The goal of the programme is to reproduce and predict, by means of suitable models, the variations of the global hydrological regime and its impact on atmospheric and oceanic dynamics, as well as variations in regional hydrological processes and water resources, and their response to change in the environment such as the increase of greenhouse gases. The GEWEX programme incorporates a major atmospheric modelling and analysis component requiring a substantial increase in computer capabilities. This is especially important because high spatial resolution is needed in climate models to achieve realistic simulations of regional effects.

Achieving the objective of GEWEX will require a major step forwards in the development of global observing techniques using new, more massive and power-demanding satellite sensors, such as lidars and radars, which will be placed on the next generation of heavy Earth-observation spacecraft in polar orbit and/or the International Manned Space Station in the late 1990s. In addition, GEWEX also requires meteorological and hydrological agencies and institutions to upgrade the collection and interpretation of essential ground-based measurements of radiation, rain, riverflow, etc.

Global Environment Monitoring System (GEMS)

The Global Environment Monitoring System (GEMS), which was established in 1975, and is administered by UNEP, is a monitoring system on many aspects of the global environment including climate and the atmosphere, long-range transport of airborne pollutants, environmental pollutants of significance to the ecosystems and human health, terrestrial renewable resources, and oceans and coastal areas. There are currently more than 20 global networks within GEMS, with activities in 142 countries. GEMS works closely with the specialized agencies of the UN, notably WHO, WMO, FAO and UNESCO.

GEMS is at the forefront of environmental work, providing through its networks and technical assessments much of the information needed for better understanding of environmental processes. Planners can use this information at all levels for appropriate management actions – ranging from conventions and protocols to desertification control measures. An important recent development is the Global Resources Information Database (GRID), which is particularly useful for studies at the national level.

global/hemispheric temperatures – the past 100 years

According to the IPCC Scientific Assessment Report, the temperature record of the past 100 years shows significant differences in behaviour between the Northern and Southern Hemispheres. A cooling of the Northern Hemisphere occurred between the 1940s and early 1970s, while Southern Hemisphere temperatures remained nearly constant from the 1940s to about 1970. Since 1970 in the Southern Hemisphere and 1975 in the Northern Hemisphere, a more general warming has been observed, concentrated into the period 1975–82, with little global warming between 1982 and 1989. However, changes of surface temperature in different regions of the two hemispheres have shown considerable contrasts for periods as long as decades throughout the past century, notably in the Northern Hemisphere.

Over periods as short as a few years, fluctuations of global or hemispheric temperatures of a few tenths of a degree are common. Some of these are related to the El Niño–Southern Oscillation phenomenon in the tropical Pacific. Evidence is also emerging of decadal timescale variability of ocean circulation and deep-ocean heat content that is likely to be an important factor in climate change.

Global Information and Early Warning System on Food and Agriculture (GIEWS)

GIEWS was established in 1975 at the request of the 1973 FAO Conference and the 1974 World Food Conference. The system monitors all aspects of food security on a worldwide scale, and it makes extensive use of agrometeorological and satellite-based data for monitoring food crop conditions and drought detection. Similar systems, which continuously monitor environmental conditions and their impact on food crops, have been established with the assistance of FAO in a number of African, Asian and Latin American countries. They significantly improve weather impact assessments, climate and climate-change readiness in the member countries.

Global Observing System (GOS)

GOS is an integral part of the World Weather Watch (WWW). In particular, the Climate System Monitoring Project (CSM) requires a systematic and efficient Global Observing System (GOS) Network with operational characteristics that are, in many ways, more demanding than weather-forecasting requirements. Data quality control, no missing data, complete records (more and different variables), station history information, good geographical distribution and representations, etc., are a few of the important areas where requirements are different. The GOS must serve as many applications as possible since the overall investment in the GOS is large compared to what is needed to improve its application to Climate Monitoring objectives.

The requirements of the GDPS together with those of the World Climate Programme and other WMO programmes are expected to increase. The highest priority of the GOS will be to meet global data requirements.

A specific project of GOS includes the expansion of GOS to meet observational requirements for climate monitoring. This project will develop proposals for expansion of the GOS so that it can satisfy the data needs of climate monitoring. This must be done by co-operative efforts among those responsible for WWW operations, for climate monitoring and for research, or a joint effort led by CBS but with a strong involvement of CCl, CMM, CIMO, JSC and EC-SAT, which will lead to proposals to be implemented by the WWW system.

Global Ocean Observing System (GOOS)

The Intergovernmental Oceanographic Commission (IOC) in co-operation with the World Meteorological Organization (WMO) has initiated the planning of a Global Ocean Observing System (GOOS) that will build on the existing data and results of research programmes to create an operational system. GOOS will meet the needs of global climate research, monitoring and prediction. Measurement categories that contribute to these needs include:
* upper-ocean monitoring programme to map, to 10° of latitude and longitude, the monthly mean distributions of heat and fresh water, sea level, carbon dioxide and nutrient contents, air–sea fluxes and plankton distribution;
* systematic deep-ocean measurements to map, to 10° of latitude and longitude, the monthly mean fluxes of heat, fresh water and dissolved chemicals, including carbon dioxide, transient tracers and nutrients;
* satellite observations of sea-surface temperature, wind stress, dynamic topography, precipitation, sea-ice concentration and chlorophyll content.

GOOS will evolve from the present national and international observing systems and experimental programmes. It will be based on:
* the Global Sea-Level Observing System (GLOSS) and ships of opportunity supplying data through the IGOSS;
* systematic deep hydrography as in WOCE and upper-ocean sampling as in TOGA, supported by moored and drifting instruments;
* a suite of operational satellite sensors including altimeter, scatterometer, imaging radiometers in the visible, infrared and microwave bands, and precision surface-temperature radiometers;
* new technologies such as acoustic and optical remote sensing, autonomous submersibles, automated data-retrieval systems for moored and drifting instruments, acoustic current profilers.

GOOS will also include worldwide ocean data centres designed to generate the products needed to meet GCOS objectives.

Global Ozone Observing System (GO₃OS)

In 1957 the WMO first established an international framework for standardized and co-ordinated ozone-observing projects, their research and related publications. It is from this that the Global Ozone Observing System eventually developed. Currently, the system has a worldwide network of approximately 140 monitoring stations. These ground-monitoring activities have been complemented by the implementation of remote-sensing techniques. This monitoring versatility has allowed the system to develop the only ozone-observing network which is capable of providing data on not only the horizontal distribution of ozone, but also its total atmospheric concentration and vertical distribution.

The Global Ozone Observing System provides ozone-related information to the UNEP–Global Environmental Monitoring System (GEMS). To date, it has published its monitoring results in a variety of reviews, as well as three major reports on the state of the ozone layer (published 1981, 1985 and 1988). Each report was published in co-operation with the US National Aeronautics and Space Administration (NASA), the US National Oceanographic and Atmospheric Administration (NOAA) and UNEP.

Ozone-related data are stored at the World Ozone Data Centre of the Atmospheric Environment Service in Canada. This Centre has been operational since the 1960s. Consequently, it is able to provide long-term extensive atmospheric ozone trends. The Canadian Centre publishes its findings in a bimonthly bulletin – *Ozone data for the world*.

Data gathering, retrieval and reporting procedures are standardized by the WMO, with support from the International Ozone Commission of the International Association of Meteorology and Atmospheric Physics.

global radiation
The total solar radiation coming from the Sun and all of the sky, including the direct beam from the unclouded Sun as well as the diffuse radiation arriving indirectly as scattered or reflected sunlight from the sky and clouds.

Global Resources Information Database (GRID)
Part of UNEP's Global Environment Monitoring System (GEMS) (see **Global Environment Monitoring System**).

Global Runoff Data Centre (GRDC)
Knowledge of the river discharge or streamflow is a basic information requirement for all kinds of hydrological investigations. Flow data are also needed for the development and verification of global models of atmospheric circulation. This has led to the collection of such data on a global scale, but it is evident that these data are also of great value for other purposes. In order to ensure that the data are easily obtainable, a central databank has been established. This is maintained by the Global Runoff Data Centre (GRDC) at Koblenz, Germany, which was officially inaugurated on 14 November 1988. The GRDC now operates with the support of Germany under the auspices of the WMO for the benefit of WMO Members and the international scientific community. GRDC also participates in the UNEP Global Environment Monitoring System for water quality (GEMS/WATER), by providing selected river discharge data to the GEMS/WATER databank held at the WMO Collaborating Centre for

Surface and Groundwater Quality at Environment Canada's "Canada Centre for Inland Waters" in Burlington, Ontario.

The GRDC databank currently consists of daily flows for 1,537 stations from 80 countries and of monthly flows for 1,390 stations from 110 countries.

Global Sea-level Observing System (GLOSS)

GLOSS is an international system, co-ordinated by IOC, to provide high-quality standardized sea-level data from a global network of sea-level stations. This network is to monitor changes in sea level due to global warming, ocean circulation patterns and climate variability, as well as to contribute to other international and regional research programmes and national practical applications. Seventy-nine countries participate in GLOSS and have designated national GLOSS contacts. Presently about 200 of the 306 proposed stations are operational. A selected set is connected to a global geodetic reference system. Data is submitted to the Permanent Service for Mean Sea Level, which disseminates and analyzes sea-level data, as well as to TOGA and WOCE sea-level centres. Substantial efforts are needed to establish and maintain stations in the Arctic and Antarctic, as well as in Africa and on remote islands.

Global Telecommunication System (GTS)

The main long-term objectives of the GTS of the World Weather Watch (WWW) are to implement fully an effective global telecommunication system operated by WMO Members to meet their needs for the collection and exchange of observational data and processed information within established time limits; and to utilize modern technology and international standards, as appropriate, to ensure that the GTS is operated in the most efficient and cost-effective manner.

Global Temperature–Salinity Pilot Project (GTSPP)

The Global Temperature–Salinity Pilot Project (GTSPP) is an IOC/WMO sponsored project intended as a trial system, that will provide the model for a global marine science data-exchange system that will serve the needs of present and future international science programmes. Its immediate task is to create a complete data- and information base of ocean temperature and salinity data, captured in real-time and submitted in fully processed form weeks to months later. GTSPP data is intended to support the World Climate Research Programme (WCRP), its associated programmes and all types of national requirements from fisheries operations to fundamental research. The programme will develop the existing oceanographic data-exchange mechanisms to provide the necessary timely and complete exchange, taking full advantage of technological advances, such as satellite data-transmission systems, CD-ROMs and information management systems based on artificial intelligence.

global warming
A climate change – on a global scale – caused by an enhanced green-house effect.

global warming potential (GWP)
The concept of relative global warming potential (GWP) has been developed to take into account the differing times that gases remain in the atmosphere. The global warming potential index is used to approximate the time-integrated warming effect due to the instantaneous release of a unit mass (1kg) of a given greenhouse gas in today's atmosphere, relative to that of carbon dioxide. This term thus describes the relative effectiveness of various greenhouse gases in contributing to potential global warming. However, because of large uncertainties (particularly with respect to the length of time that one unit of gas will remain in the atmosphere once released), the index is still a very imperfect tool for comparing emissions of different greenhouse gases. The relative importances will change in the future as atmospheric composition changes because, although radiative forcing increases in direct proportion to the concentration of CFCs, changes in the other greenhouse gases (particularly carbon dioxide) have an effect on forcing which is much less than proportional.

GLOSS see **Global Sea-Level Observing System**

GMCC
The geophysical monitoring of climate change.

Goddard Institute for Space Studies
The NASA-supported Goddard Institute for Space Studies located at Colombia University, New York.

GOOS see **Global Ocean Observing System**

GOS see **Global Observing System**

GNP see **gross national product**

GO$_3$OS see **Global Ozone Observing System**

grassland–livestock–climate relationships
The link between grassland production and livestock production dampens the magnification of climate changes exhibited by grassland productivity alone. The animals themselves act as a reservoir of biomass, buffering variations in much the same way that water reservoirs smooth out changes in streamflow. This capacity of animals to survive climate anomalies provides some livelihood stability to populations that rely on livestock husbandry for subsistence. The number of people relying on

such livelihood ranges, in various estimates, from 30 million to 70 million, mostly in Africa and Asia. In traditional grazing communities, which comprise the vast majority of people directly dependent on livestock production, breeds have been developed for their ability to survive periods of feed restriction rather than for their productivity under ideal conditions, much as traditional crops have been selected for stability of production rather than potential yield. Thus, relative reliability of livestock products is preferable to maximum productivity or migration when grassland resources vary. Actual production in the pastoral zones of Africa and Asia may be under 50% of genetic potential, and most of this loss is due to limitations imposed by climate variability.

grasslands and livestock production: climatic factors

The world's great grasslands – the Euro–Asian steppe, Tibet, the Sahel, the Pampas and North American Great Plains – lie in semi-arid regions subject to large variability of climate. The Earth's land surface is 23% grassland, supporting 1.3 billion livestock, and adds significantly to the world's food resources.

Grasslands not converted to cropland are mainly used for relatively low-density, wide-ranging livestock grazing. Pastoral production includes the raising of cattle, sheep or camels on natural grasslands managed chiefly through grazing pressure itself, with some possible limited treatment (e.g. noxious weed control). The influence of climate on different strategies of livestock husbandry is much less clearly understood than its influence on field-crop production. Few long-term data series exist matching climate and grassland productivity, and we know little about the impacts of different strategies of stock raising (e.g. nomadic versus commercial pasture-based) on the natural resource base. Nevertheless, climate clearly affects each of the three key linkages in livestock production systems: climate impacts on the primary productivity of grasslands, grazing-land productivity effects on animal production, and animal production influences on socio-economic conditions.

GRDC see **Global Runoff Data Centre**

greenhouse effect

A term used to describe the effect on solar and heat radiation of those trace gases within the atmosphere which are relatively transparent to incoming solar energy but are good absorbers of outgoing heat radiation from the Earth's surface and lower atmosphere. "Greenhouse" gases effectively help to trap heat within the atmosphere. This is a natural phenomenon which should not be confused with the "enhanced greenhouse effect".

The greenhouse effect is caused by gases in the atmosphere which allow the Sun's shortwave radiation to reach the Earth's surface, while they absorb, to a large degree, the longwave heat radiation from the

Earth's surface and from the atmosphere. Due to these gases' capacity to function as heat insulators, the temperature close to the Earth's surface is nearly 30°C higher than it would otherwise be (natural greenhouse effect).

The greenhouse effect causes heat retention in the lower atmosphere as a result of absorption and re-radiation by clouds and gases (e.g. water vapour, carbon dioxide, methane and chlorofluorocarbons) of longwave terrestrial radiation. Incoming, shortwave radiation, including visible light and heat, is absorbed by materials which then behave as black bodies re-radiating at longer wavelengths. Certain substances (e.g. carbon dioxide) absorb longwave radiation, are heated by it, then begin to radiate it, still as longwave radiation, in all directions, some of it downwards. It should be noted that, despite the use of the term "greenhouse effect", the heating in a real greenhouse is caused mainly by the physical obstruction of the glass, which prevents warm air from leaving and cooler air from entering.

However, the insulating effect caused by greenhouse gases, is in part analogous to that of the glass in a greenhouse (i.e. it is transparent to incoming shortwave radiation but partly opaque to re-radiated longwave radiation).

greenhouse gases

The so-called greenhouse gases include water vapour, carbon dioxide, tropospheric ozone, nitrous oxide and methane, which are transparent to solar radiation but opaque to longwave radiation. The most important greenhouse gas is water vapour, but human activities are not known to be significantly changing its distribution and concentration in the atmosphere.

The atmospheric gases that contribute to the enhanced greenhouse effect include carbon dioxide, methane, nitrous oxide, ozone, halocarbons, and precursors of any of these gases. Atmospheric gases that contribute to the natural greenhouse effect include water vapour, carbon dioxide, methane, nitrous oxide, ozone and other lesser gases. A climate convention would need to focus on those anthropogenic changes in greenhouse gases that cause an enhanced greenhouse effect.

The concentration of the important greenhouse gas carbon dioxide has already increased by 25% since the beginning of industrialization. This increase results from deforestation, causing an imbalance between the absorption and release of carbon dioxide by vegetation, and from the combustion of fossil fuels. Other greenhouse gases, also found in the atmosphere in increasing amounts, are methane, nitrogen oxides and, especially, chlorofluorocarbons. Anthropogenic emissions are but small increments on the large exchanges which occur naturally between the atmosphere and the oceans or the land, and thus relatively minor adjustments in the natural fluxes could significantly affect the future concentration of greenhouse gases.

Currently, the accumulation of greenhouse gases in the atmosphere is proceeding at a rate such that, by the middle of the next century, their combined effect will reach a level equivalent to doubling the pre-industrial concentration of carbon dioxide. Nevertheless, the change in the Earth's climate over the past hundred years has been substantially less than could be inferred directly from the change in the radiation budget at the surface. One reason for this is that the climate system contains both a "fast component" and a "slow component". The fast component is controlled by the atmospheric and upper-ocean heat engine which drives the whole Earth environment and determines the ultimate amplitude and geographical patterns of climate change. The slow component is controlled by the global ocean which sets the pace for climate change and may introduce delays of 50 years or more in the transient response of the Earth's climate to greenhouse forcing.

greenhouse gases: a comparison

Molecule for molecule, carbon dioxide is the least effective of the major greenhouse gases. Methane, by comparison, absorbs and re-radiates about 21 times as much heat energy, while nitrous oxide is about 206 times as effective. CFCs are even more powerful, with each molecule absorbing about 15,000 times more heat than a molecule of carbon dioxide.

However, the overall contribution of each gas to the greenhouse effect depends on two other factors as well. One of these is its atmospheric lifetime – the length of time it remains in the atmosphere before being destroyed by chemical reactions or absorbed into the biosphere or the oceans. For example, the effect of methane relative to other greenhouse gases is diminished to some extent because its atmospheric lifetime is relatively short. With longer-lived gases such as nitrous oxide and CFCs, however, the withdrawal rate is significantly lower, and new releases have a greater cumulative effect. Although accurate comparisons are difficult, it is likely that over the long term each tonne of CFC gas released into the atmosphere will have several thousand times the warming effect of the same amount of carbon dioxide. The effect of nitrous oxide can be expected to be several hundred times greater, unit for unit, while that of methane will probably be 15 to 25 times greater.

The other major factor in such comparisons is the amount of each gas that is added to the atmosphere. Since far more carbon dioxide is released into the atmosphere than any other greenhouse gas, it remains the most important single contributor to the enhancement of the greenhouse effect. It is estimated that carbon dioxide accounts for about 56% of the past decade's increase in global warming potential.

GRID see Global Resources Information Database

gross domestic product (GDP)

The value of all goods and services produced within a nation in a period of time (usually one year) charged at market prices and including taxes on expenditure, with subsidies treated as negative taxes. Essentially it is a measure of national income.

gross national product (GNP)
The total monetary value of all goods and services produced within a country during a period of time (usually one year).

groundwater
The supply of fresh water found beneath the surface of the Earth (usually in aquifers) that often supplies wells and springs.

growing degree-day
Defined as the number of degrees by which the average daily temperature exceeds a threshold or base temperature. The number of growing degree-days in a season is the summation of the growing degree-days for all days. Typical thresholds are 5°C, 10°C and 15°C.

growing season
The period of the year during which plant growth can proceed without temperature restriction.

GTS see **Global Telecommunication System**

GTSPP see **Global Temperature–Salinity Pilot Project**

H

HABITAT see **United Nations Conference on Human Settlements**

Hadley cell
A direct thermally driven and zonally symmetric atmospheric circulation first proposed by George Hadley in 1735 as an explanation for the Trade Winds. It carries momentum, sensible heat, and potential heat from the tropics to the mid-latitudes. The poleward transport aloft is complemented by subsidence in the subtropical high-pressure ridge and a surface return airflow. The variability of this cell and the Walker circulation is hypothesized to be a major factor in short-term climatic change.

Hague Meeting on the Protection of the Global Atmosphere
The 24 governments represented at the meeting held in The Hague in March 1989 issued a declaration on the atmosphere and climate change calling for a strengthening of international law and the provision of

assistance to countries to ensure that their development is not inhibited by the need for higher environmental standards. The declaration stated: "The right to live is the right from which all other rights stem. Guaranteeing this right is the paramount duty of those in charge of all states throughout the world. Today the very conditions of life on our planet are threatened by the severe attacks to which the Earth's atmosphere is subjected". Furthermore, it was considered that the problem of climate change has "three salient features, namely that it is vital, urgent and global, and called not only for the implementation of existing principles but also for a new approach, through the development of new principles of international law, including new and more effective decision-making and enforcement mechanisms".

hail

Precipitation of small balls or pieces of ice (hailstones) with a diameter ranging from 5mm to 50mm or sometimes more, falling either separately or agglomerated into irregular lumps.

half-life

The time taken by certain materials to lose half their strength. More specifically, the time taken for half the atoms in a radioactive substance to disintegrate. For example, the half-life of DDT is 15 years and the half-life of radium is 1,580 years.

halocarbons

Halocarbons containing chlorine and bromine are, molecule for molecule, among the most potent greenhouse gases in the atmosphere. They do not occur naturally but are produced industrially in large quantities. The best known members of this group of chemicals are the chloro-fluorocarbons (CFCs), which are widely used as solvents, refrigerants, spray-can propellants and foaming agents. Also significant are the halons, bromine-based compounds used as fire-extinguishing agents.

halogens

The five non-metallic elements: fluorine, chlorine, bromine, iodine and astatine.

halons

Halons are brominated chlorofluorocarbons with an extremely high ozone-depleting potential. Halons are mainly used as fire-extinguishing agents.

HDGEC

A US-sponsored programme on the human dimensions of global environmental change.

heat balance

The balance of the gains and losses of heat for a given system for a specified period.

heat flux (or thermal flux)

The amount of heat transferred across a surface of unit area in a unit of time.

heat-island effect

A "dome" of elevated temperatures over an urban area caused by local heating, and the heat absorbed by structures and pavements.

heating degree-days

Defined as the number of degrees by which the average daily temperature is below a threshold or base temperature. The number of heating degree-days in a season is the summation of the heating degree-days for all days, and is closely associated with the heating requirements of a building over season. A typical threshold is 18°C; thus a day with a temperature of 10°C is said to have "accumulated" 8 heating degree-days.

high

A term used in meteorology to denote an area of high pressure with a closed clockwise circulation of air in the Northern Hemisphere and a closed anti-clockwise circulation in the Southern Hemisphere.

Historical Archives Climate Project

Under the auspices of WMO and UNESCO, and being designed to utilize both private and public archives to determine specific meteorological events and then to correlate these events to our current climate datasets. The concept for the project is as follows :
* a test project of limited scope and duration will be undertaken to determine the feasibility and reliability of data and information retrieved from the archives;
* the test project results will be evaluated as to the benefit they provide to the climatological community;
* if beneficial, the project will be expanded as far as is possible to national archives in all countries.

At a meeting in February 1990 involving archivists from Europe, it was agreed that a pilot project should be started. The meeting further agreed that there should first be an in-depth study of the period 1725 to 1775 for the reporting of all documents containing both direct and indirect data, and an outline study of the period 1680 to 1880 for the reporting of records containing systematic meteorological surveys, reports and measurements.

Holocene

A geological epoch which refers to the most recent subdivision of the Quaternary in which we are living, the previous subdivision being the Pleistocene. The Holocene period is approximately the time since the last major glaciation, or about the past 10,000 years.

homogeneous climatic series

A data series drawn from a single population so that statistical estimates will be valid estimates of the population parameters.

HOMS see **Hydrological Operational Multipurpose Subprogramme**

HRGC

The human response to global change.

humidex

A measure of what hot weather feels like to people. In computing the humidex value, air of a given temperature and moisture content is equated in comfort to air at a higher temperature with negligible moisture content.

humidity

The water vapour content of the air.

hurricane

A very severe tropical storm with windspeeds of 120 km/h (65 knots) or more that can be many thousands of square kilometres in size. Hurricanes originate over warm tropical seas as a small low-pressure system and usually have a life span of several days. They are known in various parts of the world as either hurricanes, typhoons, tropical cyclones, or cyclones.

HWP

The Hydrology and Water Resources Programme of WMO.

hydrocarbons

Any of a vast family of compounds containing carbon and hydrogen in various combinations; they are found especially in fossil fuels. Some of the hydrocarbon compounds are major air pollutants. When other elements such as halogens are introduced, hydrocarbons can be transformed into halogenated hydrocarbons.

hydrogeology

The branch of geology which deals with groundwater and especially its occurrence.

hydrological cycle

The process of evaporation, vertical and horizontal transport of vapour, condensation, precipitation, and the flow of water from continents to oceans. The hydrological cycle is a major factor in determining climate through its influence on surface vegetation, the clouds, snow and ice, and soil moisture. The hydrologic cycle is responsible for 25–30% of the mid-latitudes' heat transport from the equatorial to polar regions.

hydrological drought

A period of abnormally dry weather sufficiently prolonged to give rise to a shortage of water, as evidenced by below-normal streamflow and lake levels and/or the depletion of soil moisture and a lowering of groundwater levels.

Hydrological Operational Multipurpose Subprogramme (HOMS)

In 1981, as a new initiative within the Operational Hydrology Programme, WMO established HOMS, the Hydrological Operational Multipurpose Subprogramme, a technology transfer system for operational hydrology.

The objectives of HOMS include:
* Improving the quality and quantity of hydrological data available;
* Aiding in the application of appropriate hydrological technology, and in related training;
* Providing an international systematic framework for the integration of the many techniques and procedures used in the collection and processing of hydrological data for water-resource systems.

HOMS is organized as a co-operative effort of the member countries of WMO. The countries that wish to participate in HOMS do so, in the first instance, by designating a HOMS National Reference Centre (HNRC), usually in the national hydrological service. To date, 105 countries have established an HNRC. The duties of an HNRC include making available national technology to users from other countries and co-ordinating national requests for the transfer of HOMS technology. The work of HNRCs is co-ordinated through the WMO Commission for Hydrology, and support is provided by the HOMS Office in the Hydrology and Water Resources Department of the WMO Secretariat.

The technology available through HOMS is presented as "components" in the form of computer software, technical or general guidance manuals, or instrument descriptions. Each component is self-contained and able to work on its own to provide the solution to some hydrological problem. At present nearly 400 HOMS components have been made available by 35 different countries. To date, over 2,000 requests for the transfer of a component had been notified to the HOMS Office, with requests running at about 350 per year. These requests came from some 86 different countries, mostly but not entirely developing countries.

hydrology
The science dealing with the properties, distribution and circulation of water.

Hydrology and Water Resources Programme of WMO (HWP)
The Hydrology and Water Resources Programme of WMO (HWP) is concerned with the quantitative and qualitative assessments and forecasts of water resources; standardization of all aspects of hydrological observations; and the organized transfer of hydrological techniques and methodology in many areas, including forecasts and mitigation of floods related to tropical cyclones, severe storms and rapid snowmelts.

hydrometric station
A station at which data on water in rivers, lakes or reservoirs are obtained on one or more of the following elements: stage, streamflow, sediment transport and deposition, water temperature and other physical properties of water, characteristics of ice cover and chemical properties of water.

hydrosphere
That part of the Earth which is composed of water; that is, the oceans, seas, ice caps, glaciers, lakes, rivers, underground water, etc.

hygrogram
The record made by a hygrograph.

hygrograph
An hygrometer which includes an arrangement for the time recording of atmospheric humidity.

hygrometer
An instrument used to measure the humidity of the air.

hypsithermal period
The period about 4,000 to 8,000 years ago when the Earth was apparently a few degrees warmer than it is now. More rainfall occurred in most of the subtropical desert regions and less in the central Midwest United States and Scandinavia. It is also called the "altithermal period" and could serve as a past climate analogue for predicting the regional pattern of climate change should the mean Earth-surface temperature increase as a result of an increase in atmospheric carbon dioxide concentration.

I

IAEA see **International Atomic Energy Agency**

IAU see **International Astronomical Union**

IBN see **International Biosciences Networks**

IBP see **International Biological Programme**

ice age

Ice ages are substantial features of the climate of the Earth. An ice age is a period in the Earth's history when ice spread towards the Equator, accompanied by a general lowering of surface temperatures, especially in temperate latitudes. In the most recent ice age in the Pleistocene period, which ended about 10,000 years ago, there were at least four major ice advances, with the margin reaching about 52°N over northwest Europe and about 45°N in northeast United States. With this change in location of ice surfaces the whole atmospheric circulation altered, the main climatic belts being compressed and pushed towards the Equator. At present we are in an interglacial circulation.

The cause of ice ages is uncertain, but Milankovitch proposed that variations in the Earth's orbit and the inclination of its axis caused the ice ages. It is generally agreed that this is the likely trigger mechanism for ice ages but that feedback or other processes must also be involved. Alternative hypotheses are based on changes in solar activity. There have been many ice ages in Earth history, dating back as far as the Precambrian in which at least 15 major groups of ice ages occurred in the Laurasian and Gondwanaland continents. Further groups occurred during the Carboniferous, with glaciation confined to the southern hemisphere. The third main group of ice ages occurred during the Quaternary, the most recent being the Pleistocene ice age as noted above.

Between 1550 and 1850 temperatures in much of the Northern Hemisphere fell to their lowest since the last ice age, and this period has been called the Little Ice Age. At that time, alpine glaciers advanced, and settlement in many northern areas, such as Greenland, Iceland and Northern Norway, had to be abandoned.

ice and snow albedo

The reflectivity of ice- and snow-covered surfaces. The albedo of freshly fallen snow may be as much as 90%, while older snow may have values of 75% or less. The larger the areal extent of snow and ice cover, the higher the albedo value. The surface albedo will also increase as a function of the depth of snow cover up to 13cm and be unaffected by increased snow cover after reaching that depth.

iceberg

A mass of land ice that has broken away from land and floats in the sea, or becomes stranded in shallow water. The unmodified term "iceberg" usually refers to the irregular masses of ice formed by the calving of glaciers along an orographically rough coast.

ice front

The floating vertical cliff that forms the seaward face or edge of a glacier or an ice shell that enters water. It can vary from 2m to 50m in height.

ice pellets

Precipitation of transparent or translucent pellets of ice, which are spherical or irregular, and which have a diameter of 5mm or less.

ice sheets

The largest form of glaciers, ice sheets cover extensive areas and are often thick enough to bury all but the highest peaks of entire mountain ranges. During the Pleistocene epoch, ice sheets covered large parts of North America and northern Europe but they are now confined to polar regions (e.g. Greenland and Antarctica). Almost all of Antarctica is covered by ice that locally is 2,500m thick and the Greenland ice sheet is more than 3,000m thick. Smaller ice sheets occur in Iceland, Spitsbergen, and other Arctic islands.

ice shelf

A sheet of very thick ice attached to the land on one side, but most of it floating. On the seaward side, it is bounded by a steep cliff (ice front) 2–50m or more above sea level. Ice shelves have formed along polar coasts (e.g. Antarctica and Greenland); they are very wide with some extending several hundreds of kilometres towards the sea from the coastline. They increase in size from annual snow accumulation and seaward extension of land glaciers, and decrease in size as a result of warming, melting and calving.

Icelandic low

The low-pressure centre located near Iceland (mainly between Iceland and southern Greenland) on mean charts of sea-level pressure. It is a principal centre of action in the atmospheric circulation of the mid- and northern Northern Hemisphere and is most intense during winter.

ICID

The International Commission on Irrigation and Drainage.

ICL see **Inter-Union Commission on the Lithosphere**

ICOLP
An industry co-operative programme for ozone layer protection which deals with information supply, new products, conferences, technology transfer seminars, legislation and key contracts within industry and government. ICOLP runs OZONET (an ozone network).

ICSTI see **International Council for Scientific and Technical Information**

ICSU see **International Council of Scientific Unions**

IDNDR see **International Decade for Natural Disaster Reduction**

IEA
The International Energy Agency.

IFAD
The International Fund for Agricultural Development.

IGAC see **International Global Atmospheric Chemistry Project**

IGAP see **International Global Aerosol Programme**

IGBP see **International Geosphere–Biosphere Programme**

IGFA
International Group of Funding Agencies for Global Climate Change.

IGOSS see **Integrated Global Ocean Services System**

IGU see **International Geographical Union**

IGY see **International Geophysical Year**

IHD see **International Hydrological Decade**

IHP see **International Hydrological Programme**

IIASA see **International Institute for Applied Systems Analysis**

IJC see **International Joint Commission**

ILO
The International Labour Organization.

IMCO see **Intergovernmental Maritime Consultative Organization**

IMF see **International Monetary Fund**

IMO see **International Meteorological Organization**

impact of rising sea level

Rising sea level could have far-reaching consequences. The effects upon tropical, lesser-developed countries with high coastal population densities could be devastating, and costly engineering solutions may not be feasible. Global long-term sea-level changes during the next few decades will be less important than effects which take place locally on short timescales as a consequence of tides, waves, storm surges, and seasonal cycles. Higher relative mean sea level may cause more frequent flooding associated with a combination of these local events. As time goes by, even the lesser storms would cause destructive floods. Thus, higher relative mean sea level will most likely not be perceived by the public as a gradual rise, but as a series of more frequent extraordinary events. Some potential impacts of rising relative mean sea level are accelerated erosion of beaches and coastal shores, salt-water contamination of coastal aquifers, salt-water intrusion a greater distance landward into estuaries and bays, increased frequency of coastal flooding, and the destruction of coastal terrestrial vegetation and replacement by salt-tolerant species.

implementing authority

Any governmental agency at any level having appropriate authority to authorize and execute the implementation of any particular action and the jurisdiction to enforce an action.

INC see **Intergovernmental Negotiating Committee on a Framework Convention on Climate Change**

INC/FCCC

The International Negotiating Committee, Framework Convention on Climate Change.

Indian summer

A period, in mid- or late autumn, of abnormally warm weather, generally clear skies, sunny but hazy days, and cool nights. The term is most often used in the northeastern United States, but its usage extends throughout English-speaking countries. It dates back at least to 1778, but its origin is not certain; the most probable suggestions relate it to the way that the American Indians availed themselves of this extra opportunity to increase their winter stores.

INFOCLIMA see **Climate Data Information Referral System**

infrared radiation

Electromagnetic radiation in the approximate wavelength range from about $0.7-1,000\mu$m; 52% of the total solar radiation intensity is contained within this range of wavelengths.

INQUA see **International Union for Quaternary Research**

insolation

The intensity of either direct or global (direct and diffuse) solar radiation on a unit area at a specified time on a specified surface. Its value is dependent upon the solar constant, the time of year, the geographical latitude of the receiving surface, the slope and aspect of the surface, and the transparency of the atmosphere. The latitudinal variation in the insolation supplies the energy for the general circulation of the atmosphere.

INSULA see **International Scientific Council for Island Development**

Integrated Global Ocean Services System (IGOSS)

The operational network for the global collection and exchange of oceanographic (surface and subsurface temperature, salinity, currents) data in real time, using the GTS. IGOSS was established in 1967 jointly by the IOC and the WMO, and consists of national facilities and services provided by member states. The purpose of IGOSS is to make available to member states data and information required to provide efficient and effective ocean services for both operational and research applications. IGOSS promotes, co-ordinates and develops the international arrangements necessary for the timely acquisition and exchange of data, the provision of services, and the dissemination of products in the form of observations, analyses and forecasts. Approximately 40,000 subsurface observations and 5,000 salinity observations are transmitted annually. The IGOSS Ship-of-Opportunity Programme includes over 200 ships taking observations, many of which carry automated systems for automatically encoding and transmitting data via satellite. Substantial national commitments are needed to provide adequate coverage, particularly in such data-sparse areas as the Indian and Southern Oceans.

interception

A term used to describe the process by which precipitation is caught and retained on vegetation without reaching the ground.

interglacials

The periods between the glaciations of an ice age.

Intergovernmental Maritime Consultative Organization (IMCO)

The international body that regulates many aspects of the operation of ships on the high seas, including the pollution of the sea.

Intergovernmental Negotiating Committee (INC) on a Framework Convention on Climate Change

In December 1990, the UN General Assembly decided to establish a single intergovernmental negotiating process under the auspices of the General Assembly, supported by the United Nations Environment Programme and the World Meteorological Organization, for the preparation of an effective framework convention on climate change. The General Assembly stated that such a framework convention should contain appropriate commitments, and any related instruments as might be agreed upon, taking into account proposals that may be submitted by states participating in the negotiating process, the work of the Intergovernmental Panel on Climate Change and the results achieved at international meetings on the subject, including the Second World Climate Conference. The General Assembly further decided that the first negotiating session should be held at Washington DC, in February 1991 and that further meetings should be held in May/June 1991, September 1991, and November/December 1991 and, as appropriate, between January and June 1992.

The General Assembly also considered that the negotiations for the preparation of an effective framework convention on climate change, containing appropriate commitments, and related legal instruments as might be agreed upon, should be completed prior to the United Nations Conference on Environment and Development in June 1992 and opened for signature during the Conference.

Intergovernmental Oceanographic Commission (IOC) of UNESCO

A collaborating partner in the World Climate Programme, the IOC has several specific areas of interest including:

* co-ordination of the Global Sea Level Monitoring System (GLOSS) which monitors and analyzes sea-level data;
* work on marine science and ocean services for development, together with a major assistance programme to enhance the marine science capabilities of developing countries.

Intergovernmental Panel on Climate Change (IPCC)

The idea of the IPCC was first suggested in 1987 and endorsed by the WMO Executive Council and the UNEP Governing Council at their meetings in 1988. The IPCC was established to report to the governing bodies of WMO and UNEP with three specific tasks: to assess the scientific information related to the various aspects of the climate change issue, to evaluate the environmental and socio-economic impacts of climate change, and to formulate realistic response strategies for the management of the greenhouse issue.

The first session of the IPCC held in Geneva in November 1988 established three working groups:

– Working Group I under the Chairmanship of Dr J. T. Houghton

(United Kingdom): Responsible for assessing all available scientific information on factors affecting climate change including greenhouse gases, responses to these factors of the atmosphere–ocean–land–ice system, assessment of current capabilities of modelling global and regional climate change and their predictability, past climate records and presently observed climate anomalies, projections of future climate and sea level, and the timing of changes. Its report identifies the range of projections and their regional variations, gaps and uncertainties.

- Working Group II under the Chairmanship of Professor Yu. Izrael (USSR): Responsible for reviewing environmental and socio-economic impacts of climate change in an integrated manner. The Group also addressed the evaluation of impacts on a regional and national scale of climate warming and sea-level rise, on agriculture, forestry, health, energy and water resources; with special regard to floods, droughts and desertification.
- Working Group III originally under the Chairmanship of Dr F. Bernthal (USA): Concerned with the policy dimension and, in particular, future emissions of greenhouse gases, impacts of changing technology, sources and sinks, adaptation to climate change, strategies to control or reduce emissions and other human activities that may have an impact on climate (e.g. deforestation, changing land-use) and their social and economic implications, including legal matters.

To oversee the work of the IPCC an intergovernmental bureau was established, chaired by Professor B. Bolin of Sweden, and a Secretariat located in the WMO.

At its second session in Nairobi, in June 1989, the IPCC set up a Special Committee on Matters related to Developing Countries. The Working Groups completed their draft reports in May/June 1990, together with draft 20-page summaries for policy-makers. The IPCC held its third session in Washington DC in February 1990, and its fourth session, which approved the First Assessment Report, in Sundsvall, Sweden, in August, 1990. (See also **IPCC** et seq.)

International Astronomical Union (IAU)

Founded in 1919 to provide a forum where astronomers from all over the world could develop astronomy in all its aspects through international co-operation. Since that time the Union has devoted itself to this purpose and its reputation is such that most of the international co-operation in astronomy is conducted through the IAU. The Union is very diverse in character and its scientific activity is reflected in the work of its 40 Commissions, which refer to all phenomena in outer space.

International Atomic Energy Agency (IAEA)

The United Nations agency concerned with all aspects of the peaceful use of atomic energy and the commercial and scientific uses of radioisotopes. IAEA has its headquarters in Vienna.

International Bank for Reconstruction and Development (World Bank)

An international bank formed as a result of the 1945 United Nations Monetary and Financial Conference held at Bretton Woods, New Hampshire, to facilitate trade and development. The Bank now has 117 members, and works in association with other agencies, including development banks for Africa, Asia and Latin America, and the International Finance Corporation (IFC), the International Development Association (IDA).

International Biological Programme (IBP)

A world study – covering 1964–74 – of biological productivity and human adaptability initiated by the ICSU. The study was divided into seven sections: productivity of terrestrial communities, production processes on land and in water, conservation of terrestrial communities, productivity of freshwater communities, productivity of marine communities, human adaptability, and the use and management of biological resources.

International Biosciences Networks (IBN)

A Joint ICSU–UNESCO programme established by the Bio-Unions of ICSU and UNESCO in 1979 to assist developing countries build up their capacities in the biosciences. The IBN is overseen by an International Steering Committee and its regional activities are co-ordinated by Regional Networks.

International Centre for Earth Tides   see **Federation of Astronomical and Geophysical Data Analysis Services**

International Conference on Water and the Environment

Held in Dublin, Ireland, from 26–31 January 1992, the conference was hosted by the Government of Ireland and convened by the World Meteorological Organization (WMO) on behalf of the United Nations Administrative Committee on Co-ordination Inter-Secretariat Group for Water Resources (ACC/ISGWR). The Conference addressed critical freshwater issues and related development issues for the 21st century. It acted as a formal entry for these issues into the United Nations Conference on Environment and Development (UNCED) held in Rio de Janeiro, Brazil, in June 1992.

International Council for Scientific and Technical Information (ICSTI)

ICSTI was established in June 1984 as the successor to the ICSU Abstracting Board. The purpose of ICSTI is to increase accessibility to, and awareness of, scientific and technical information. ICSTI aims to foster communication and interaction among all participants in the information transfer chain, in order to develop appropriate tools to meet better the information requirements of the world community of scientists and technologists.

International Council of Scientific Unions (ICSU)

The ICSU was created in 1931 to promote international scientific activity in the different branches of science and their applications for the benefit of humanity. ICSU encourages the exchange of scientific information, initiates programmes requiring international scientific co-operation, and studies and reports on matters relating to the social and political responsibility and treatment of scientists. Since its creation, it has vigorously pursued a policy of non-discrimination, affirming the rights and freedom of scientists throughout the world to engage in international scientific activity without regard to such factors as citizenship, religion, creed, political stance, ethnic origin, race, colour, language, age or sex.

ICSU is a non-governmental organization with two categories of membership: scientific academies or research councils which are national multidisciplinary bodies (76 members), and scientific unions, which are international disciplinary organizations (20 members). The complement of these two groups provides a wide spectrum of scientific expertise, enabling members to address major international, interdisciplinary issues which none of them could handle alone. In addition, ICSU has 26 Scientific Associates.

The Council seeks to accomplish its rôle in several ways. First, it initiates, designs and co-ordinates major international interdisciplinary research programmes, such as the International Geophysical Year (1957-58), the International Biological Programme (1964-74), or the recently launched International Geosphere–Biosphere Programme: A Study of Global Change (IGBP), which complements the joint WMO/ICSU World Climate Research Programme, and aims to describe and understand the interactive physical, chemical and biological aspects of the total Earth system. Secondly, ICSU creates interdisciplinary bodies which undertake activities and research programmes of interest to several member bodies. Examples of such activities include Antarctic, oceanic, space and water research, problems of the environment, genetic experimentation, solar–terrestrial physics, and biotechnology.

In addition to these programmes and activities, which seek to break the barriers of specialization, several bodies set up within ICSU address matters of common concern to all scientists, such as the teaching of science, data, science & technology in developing countries, ethics, and the free circulation of scientists.

International Decade for Natural Disaster Reduction (IDNDR)

In an attempt to promote international co-operation on natural disasters, the IDNDR was launched by the United Nations on 1 January 1990. This was the first globally concerted effort to reduce the impact of natural hazards in all their forms. The aim of the Decade, set out in UN Resolution 42/169, is to reduce through concerted international action, especially in developing countries, the loss of life, property damage, and social and economic disruption caused by natural disasters, such as earthquakes, windstorms, tsunamis, floods, landslides, volcanic eruptions, wildfires, grasshopper and locust infestations, drought and desertification, and other calamities of natural origin. The goals are:

* to improve the capacity of each country to mitigate the effects of natural disasters expeditiously and effectively, paying special attention to assisting developing countries in the assessment of disaster damage potential and in the establishment of early warning systems and disaster-resistant structures when and where needed;
* to devise appropriate guidelines and strategies for applying existing scientific and technical knowledge, taking into account the cultural and economic diversity among nations;
* to foster scientific and engineering endeavours aimed at closing critical gaps in knowledge in order to reduce loss of life and property;
* to disseminate existing and new technical information related to measures for the assessment, prediction and mitigation of natural disasters;
* to develop measures for the assessment, prediction, prevention and mitigation of natural disasters through programmes of technical assistance and technology transfer, demonstration projects, and education and training, tailored to specific disasters and locations, and to evaluate the effectiveness of these programmes.

International Earth Rotation Service see Federation of Astronomical and Geophysical Data Analysis Services

International Energy Agency (IEA)

An autonomous body which was established in November 1974 within the framework of the Organisation for Economic Co-operation and Development (OECD) to implement an international energy programme. It carries out a comprehensive programme of energy co-operation among twenty-one of the OECD's 24 member countries. The basic aims of the IEA are:

* co-operation among IEA participating countries to reduce excessive dependence on oil through energy conservation, development of alternative energy sources, and energy research and development;
* an information system on the international oil market as well as consultation with oil companies;
* co-operation with oil-producing and other oil-consuming countries with a view to developing a stable international energy trade as well as the rational management and use of world energy resources in the interest of all countries;

* a plan to prepare participating countries against the risk of a major disruption of oil supplies and to share available oil in the event of an emergency.

International Federation for Information and Documentation (IFID)

The Federation was founded in 1895. Its aim is to promote, through international co-operation, research in and development of all aspects of the organization, storage, retrieval, dissemination and evaluation of information, however recorded, in the fields of science, technology, social sciences, arts and the humanities.

International Federation for Information Processing (IFIP)

The IFIP began its official existence in 1960, under the auspices of UNESCO. Its basic aims are: to promote information science and technology; to advance international co-operation in the field of information processing; to stimulate research, development and application of information processing in science and human activity; to further the dissemination and exchange of information on information processing; and to encourage education in information processing.

International Geographical Union (IGU)

Since its creation in Brussels in 1922, the Union has consisted of three major components: a General Assembly of the delegates appointed by the member countries, an Executive Committee, as well as Commissions and Study Groups which continue their work between General Assemblies. The IGU organizes international congresses, regional conferences and specialized symposia. The first International Geographical Congress was held in Antwerp in 1871, and subsequent meetings led to the establishment of a permanent organization. The congresses are now held every four years, most recently in Sydney in 1988 and in Washington DC in 1992. Among other things, IGU promotes the study of geographical problems, initiates and co-ordinates geographical research requiring international co-operation, provides for the participation of geographers in the work of relevant international organizations, and facilitates the collection and diffusion of geographical data and documentation in and between all member countries.

International Geosphere–Biosphere Programme (IGBP)

The IGBP programme of the International Council of Scientific Unions (ICSU) was initiated in 1986 to "describe and understand the interactive physical, chemical and biological processes that regulate the total Earth system . . . the changes that are occurring in this system, and the manner in which they are influenced by human actions". This programme builds upon and integrates other scientific work by ICSU, especially through its Scientific Committee on Problems of the Environment (SCOPE), and the WCRP. IGBP emphasizes the biochemical cycles and the WCRP the physical climate system. The programme was initiated by ICSU in order to examine the links between the biological, chemical and physical processes in the whole Earth system and the interactions between the different parts of the system. The IGBP provides a framework for a wide range of interdisciplinary programmes. It is being developed in close co-operation with other international scientific programmes, some of which have been identified as core projects in relation to studies of global change. These programmes include IGACP, JGOFS, PAGES, BAHC, and GCTE.

The first meeting of the Special Committee of IGBP which took place in July 1987 identified several special themes that will guide the IGBP's future work, including:
* documenting and predicting climate change;
* observing and improving understanding of dominant forcing functions;
* improving understanding of interactive phenomena in the total Earth system; and
* assessing effects of global change that have major implications for renewable and non-renewable resources.

International Geophysical Year (IGY)
The IGY of 1957–58 covered the observational period from 1 July 1957 to 31 December 1958. Enhanced observational programmes were carried out on "Regular World Days" (three days at every new moon) and on "Special World Days", selected in accordance with solar activity. The meteorological programme, besides the usual observations, included measurements of solar radiation and atmospheric ozone. Although ozone measurements with the Dobson spectrophotometer began 25 years earlier, the internationally co-ordinated observations of ozone started during the IGY. The Mauna Loa Observatory in Hawaii also started CO_2 measurements in 1958 under the IGY.

A new feature of the IGY was the high-altitude rocket observation programme with the co-operation of the USA, USSR, UK and Japan and the launching of the first satellites by the USSR and the USA.

Some of the stations set up in Antarctica for the IGY have become permanent, and the formulation of a scientific co-operation programme in the Antarctic led to the signing of the Antarctic Treaty.

International Global Aerosol Programme (IGAP)
The IGAP plan resulted from the deliberations of the Joint Working Group and seven working subgroups of the International Aerosol Climatology Project (IACP). The Joint Working Group on IACP was formed under the auspices of the Commissions on Radiation (IRC), Clouds and Precipitation (ICCP) and Atmospheric Chemistry and Global Pollution (ICACGP) of the International Association for Meteorology and Atmospheric Physics (IAMAP). Over 80 scientists contributed to the formulation of the IGAP.

The overall objective of the International Global Aerosol Programme is to improve understanding of the rôle of atmospheric aerosols in the forcing mechanisms and forecasting of changes in global climate and in geospheric-biospheric processes. Progress towards this objective will depend on the development of a more complete understanding of the source—transformation—transport—sink cycles of the aerosols, and of their properties, concentrations, and interactions with other components of the atmosphere, the biosphere, and of the geosphere.

The activities proposed within the plan for the International Global Aerosol Programme can be grouped into the following broad categories:
* global aerosol climatologies and model descriptions;
* monitoring of global aerosol properties, distributions, and effects;
* process studies.

International Global Atmospheric Chemistry Project (IGAC)

A core project of IGBP, jointly organized with the ICSU/IUGG/IAMAP Commission on Atmospheric Chemistry and Global Pollution. Its goal is to document and understand the processes regulating biogeochemical interactions between the terrestrial and marine components of the biosphere and the atmosphere, and their rôle in climate. It consists of several research projects, which address natural variability and anthropogenic perturbations in the composition of the atmosphere over terrestrial tropical, polar and mid-latitude regions, as well as over the oceans. Other research efforts will address, through observations and modelling, the global distribution of chemically and radiatively important species (including emission rates and other processes governing their abundances) and the rôles of these substances in cloud condensation. The results will yield a marked improvement in our understanding of the processes responsible for regulating the abundance of atmospheric constituents that are of relevance to climate.

international greenhouse gas treaties

At present there exists no specific international treaty which seeks to regulate comprehensively the emission into the atmosphere of greenhouse gases (such as the 1982 UN Convention on the Law of the Sea in relation to pollution of the seas). There is also no treaty establishing the liability of one state to repair damage in another, or to the global commons (i.e. the area beyond national jurisdiction), or for the consequences of greenhouse gas emissions emanating from areas within its jurisdiction. Accordingly, in the event that greenhouse gases lead to climate change and sea level adversely affects any state or territory, that state and/or the international community as a whole would - under present conditions - have difficulty in identifying a specific treaty obligation, or breach, enabling it to base a claim for preventive measures or reparation.

The 1985 Vienna Ozone Convention and the 1987 Montreal Protocol, as amended, have an effect on certain ozone-depleting substances.

However, their effect is limited to the extent that it does not regulate the whole cycle of production, and it also does not establish any rules to limit the emission into the atmosphere of any existing ozone-depleting substances. The 1979 UN ECE Convention on Long-Range Transboundary Air Pollution (the first "environmental" treaty involving almost all nations of East and West Europe, the USA, and the USSR), establishes a qualified general obligation on "state parties" to "endeavour" to limit, reduce and prevent air pollution. However, it does not contain a rule on state liability as to damage. Determining the extent of individual states' obligations in respect of transboundary and global environmental pollution is relatively limited.

The International Court of Justice has stated that the principle of sovereignty embodies "the obligation of every state not to allow its territory to be used for acts contrary to the rights of other states", and in the much cited "Trail Smelter" case an ad hoc arbitral tribunal held that: "Under the principles of international law . . . no state has the right to use or permit the use of territory in such a manner as to cause injury by fumes in or to the territory of another of the properties or persons therein, when the case is of serious consequence and the injury is established by clear and convincing evidence." Most writers have accepted this formulation as a rule of customary international law. More recently, almost all states in the international community have accepted Principle 21 of the 1972 Stockholm Declaration and Article 30 of the 1974 Charter of Economic Rights and Duties of States.

Taken together, the principle of state sovereignty the "Trail Smelter" case, Principle 21 of the Stockholm Declaration, the 1979 UN ECE Convention, the 1982 UN Convention on the Law of the Sea, and the Charter of Economic Rights and Duties of States, suggest that the following rule of customary international law has emerged: "International law imposes upon states an obligation not to emit any substance into the atmosphere to a level capable of causing harm, or likely to cause harm, to people, property and the environment, to states, other territories and the global commons."

International Hydrological Decade (IHD)
A ten-year programme of international scientific co-operation in research on water problems, planned and directed by UNESCO, which ran from January 1965.

International Hydrological Programme (IHP)
The UNESCO-based IHP addresses itself to the long-term goal of advancing our understanding of the processes occurring in the water cycle and integrating this knowledge into water resources management. It aims to improve scientific and technical knowledge of freshwater processes, to train the necessary personnel, to build up research and training institutions and to stimulate conservation and responsible management of water resources by decision-makers and the public.

Scientific understanding of interactions between land, vegetation, oceans, atmosphere and human actions is one of the IHP's priorities. By definition, the problems are multidisciplinary, and the IHP takes an integrated approach in studying the various links and linkages that make up the world water cycle.

IHP wants to see the application of hydrological research results to integrated water management strategies. These include: improving agricultural productivity, providing water for irrigation and people, controlling urban water problems, and developing land-use practices that meet these needs while still reducing flood damage and the degradation of soils and water. It is expected that good water management in the humid tropics will also bring benefits to areas outside the region.

International Institute for Applied Systems Analysis (IIASA)

IIASA is a non-governmental research institute sponsored by scientific organizations from East and West. It brings together scientists from many nations and a variety of disciplines. Its purpose is to develop practical options to deal with issues of international importance through the application of system sciences. The Institute's effectiveness is rooted in its international sponsorship and focus, its nonpolitical status, its freedom to choose its research agenda from a variety of pressing international issues, its interdisciplinary base, and its worldwide network of collaborating organizations. Recent projects have included studies on global climate changes, world agricultural potential, energy resources, acid rain, computer-integrated manufacturing, the social and economic impacts of demographic changes, and the theory and methods of systems analysis. The basis of IIASA's scientific research is the development and use of computer models to help define how global issues and problems may evolve in the future. The objective is to develop viable policy options that can be implemented through international co-operation. IIASA was founded in 1972, on the initiative of the USA and the USSR, with the eventual participation of another 13 countries in the East and West. IIASA has member organizations in the following countries: Austria, Bulgaria, Canada, Czechoslovakia, Finland, France, Germany, Hungary, Italy, Japan, the Netherlands, Poland, Sweden, Russia and the USA.

International Joint Commission (IJC)

A bi-national Commission between the USA and Canada created under the 1909 Boundary Water Treaty. The IJC has three primary functions:
* quasi-judicial with responsibility for approving applications to affect natural flows or levels of boundary waters;
* investigation of matters at the request of the two governments with the limitation that the resulting recommendations are not binding on the governments and can be modified or ignored;
* surveillance/co-ordination of the implementation of recommendations at the request of the governments.

International Meteorological Conference (1853)
The first International Meteorological Conference was held in Brussels in August, 1853. The main achievements of the Brussels Conference were: the adoption of a standard form of ship's log, a set of standard instructions for meteorological observations at sea, and a system for the collection of ships' logs.

International Meteorological Congress (1873)
In Vienna, in September 1873, the First International Meteorological Congress created the non-governmental International Meteorological Organization (IMO) which was the predecessor of WMO. Among other things, the Vienna Congress adopted definitions of meteorological phenomena and a list of symbols to be used in climatological tables and on weather charts. It stressed the necessity for global observations. Subsequently, the meteorologists organized a programme on further standardization of meteorological instruments and methods of observation, and they also took the first steps towards defining an international telegraphic code for the exchange of observational data.

International Meteorological Organization (IMO)   see World Meteorological Organization: early history

International Monetary Fund (IMF)
The IMF was founded with the aim of reforming and stabilizing currencies as a result of the 1945 United Nations Monetary and Financial Conference at Bretton Woods, New Hampshire.

International Oceanographic Data Exchange (IODE)
This IOC programme maintains procedures and fosters facilities for the international exchange of ocean data through a network of national oceanographic data centres, designated national agencies, responsible national oceanographic data centres, and world centres. A total of 42 countries now have national data centres or designated national agencies. The marine database of the World Data Centre system contains data from more than 2,250,000 observations, biological and marine geological data, and current measurements. The IOC has also developed a Marine Environmental Data and Information Referral System (MEDI), a data source guide which describes data sets available at data centres. Actions are being taken by IOC and WMO, through IGOSS and IODE, to create a timely and complete global ocean data and information base to support the World Climate Research Programme and national programmes.

International Polar Year
The years 1882 and 1932, during which participating nations undertook increased observations of geophysical phenomena in polar regions. The

observations were largely meteorological, but included auroral and magnetic studies.

International Satellite Cloud Climatology Project (ISCCP)

The ISCCP, is one of several systematic data-collection and analysis efforts organized by the WCRP to derive essential climatological information. The objectives of the ISCCP, initiated on 1 July 1983, is to produce a continuous record of monthly mean cloud amounts and optical properties over the whole Earth. The properties of the cloud fields are computed every month for each 2.5° latitude by 2.5° longitude box. To produce this information with appropriate time and space sampling, visible and infrared image data is collected, produced every three hours by a multiple satellite array distributed around the Equator. The vast task of assembling and processing these data from the different satellite systems, and extracting meaningful cloud parameters, is carried out by centres in several countries, before being finally quality controlled and archived by the NASA Goddard Institute for Space Studies.

International Scientific Council for Island Development (INSULA)

In November 1989, INSULA, a new international non-governmental organization, was created whose aim is to promote the sustainable development of small islands in all regions of the world. The need for such an organization had been recognized by international UNESCO MAB meetings during the period 1986 to 1989.

Among the Council's objectives is to encourage technical, scientific and cultural co-operation in assisting island communities in integrated planning and the judicious use of their natural and human resources. The action of INSULA will be essentially catalytic, designed to promote the application to the specificities of small island situations of multidisciplinary scientific research and technology and innovations in education, culture and communications.

International Service of Geomagnetic Indices see Federation of Astronomical and Geophysical Data Analysis Services

International Society of Soil Science (ISSS)

Founded in 1924, the ISSS comprises 7,500 individual members and 60 affiliated national soil science societies. The aim of the society is to foster all branches of soil science and its application.

International Space Year (ISY)

The projects in the International Space Year are: Polar Stratospheric Ozone (PSO), Greenhouse Effect Detection Experiment (GEDEX), and Polar Ice Extent (PIE).

The ISY concept is intended to enhance international collaboration among space agencies and scientists on the global phenomena of the Earth's environment, as well as to enhance public awareness of the benefits deriving from space activities, in particular from space-based Earth observation, for scientific and applications research. The concept was initiated by the United States in 1985 and endorsed by the UN in 1989. The year 1992 was chosen for ISY because it is the 35th anniversary of the launch of Sputnik and the 500th anniversary of the discovery of the New World by Columbus, among other notable events. ISY activities are co-ordinated by the Space Agency Forum for the ISY (SAFISY), which represents 28 industrialized and developing countries and international organizations committed to participate in ISY.

International Tropical Timber Agreement (ITTA)
The ITTA came into being in 1983. It brings together tropical timber exporting and importing countries. The principal purpose of the Agreement is to promote trade in tropical timber and wood processing in the producing countries, combined with measures to protect the tropical forests and promote sustainable production of tropical timber.

International Tropical Timber Organization
The International Tropical Timber Organization (ITTO) is a forum for co-operation among producers and consumers of tropical timber. Over 40 countries belong to it, representing more than 75% of the existing tropical forests and more than 95% of international trade in tropical timber. The organization has its headquarters in Yokohama, Japan.

International Union for Conservation of Nature and Natural Resources (IUCN)
An independent international body, founded in 1948, with headquarters at Morges, Switzerland. It promotes and initiates scientifically based conservation measures and co-operates with the United Nations and other intergovernmental agencies, and with its sister organization the World Wildlife Fund, which exists primarily to raise and allocate funds.

International Union for Physical and Engineering Sciences in Medicine (IUPESM)
The Union was founded in 1980 by the International Federation for Medical and Biological Engineering and the International Organization for Medical Physics. Its objectives are: to contribute to the advancement of medical science and technology; to organize international co-operation and promote communication among those engaged in health-care science and technology; to co-ordinate activities of mutual interest to the engineering and physical sciences within the health-care field, such as international and regional scientific conferences, seminars, working groups, regional support programmes and scientific and technical publications; and to represent the professional interests and views of

engineers and physical scientists in the health-care community.

International Union for Pure and Applied Biophysics (IUPAB)

IUPAB was established in 1961 in Stockholm, as the International Organization for Pure and Applied Biophysics. The objectives are: to organize international co-operation in biophysics and promote communication between the various branches of biophysics and allied subjects; and to encourage within each adhering body co-operation between the societies that are interested in the advancement of biophysics in all its aspects.

International Union for Quaternary Research (INQUA)

The Union was established in 1928 as the International Association for the Study of the European Quaternary. The scope was enlarged to cover the world in 1932, and the present name of the Union was adopted in 1965. Its aim is to co-ordinate research on the Quaternary era throughout the world.

International Union of Biochemistry (IUB)

The IUB, created in January 1955, is dedicated to promoting international co-operation in biochemistry via research, discussion, publication and standardization, and is committed to promoting high standards of biochemistry throughout the world. The IUB fulfils these mandates in a variety of ways. Committees, charged with sponsoring symposia, educational activities and discussion of frontier research topics, sponsor meetings in all parts of the world.

International Union of Biological Sciences (IUBS)

IUBS was created in 1919, and held its first General Assembly in that year. Its objectives are to promote the study of biological sciences, to facilitate and co-ordinate research and other scientific activities that require international co-operation, to ensure the discussion and dissemination of international conferences, and to assist in the publication of their results. The IUBS scientific programmes are of an interdisciplinary nature. They are undertaken in collaboration with the national scientific authorities and in co-operation with other international organizations, both intergovernmental and non-governmental. These programmes include, among others, the Decade of the Tropics, the biological monitoring of the state of the environment (bio-indicators), biotechnology, biological education, biological nomenclature and taxonomy, biological oceanography, and the study of fertility regulation in humans, animals and plants.

International Union of Forestry Research Organizations (IUFRO)

The IUFRO was established in 1890, and reorganized in its present form in 1971. Its main objectives are to rationalize research techniques, to standardize systems of measurement, and to promote international co-operation in scientific studies, forestry and forest products, including: site and silviculture; forest plants and forest protection; forest operations and techniques; planning, economics, growth and yield, management, and policy; forest landscape, recreation and tourism; statistical methods, mathematics and computerization; information systems and terminology; remote sensing; management of forestry research; and history.

International Union of Geodesy and Geophysics (IUGG)
Established in 1919, the IUGG is dedicated to the scientific study of the Earth and the applications of the knowledge gained by such studies to the needs of society, such as mineral resources, reduction of the effects of natural hazards, and environmental preservation. Its objectives are the promotion and co-ordination of physical, chemical and mathematical studies of the Earth and its environment in space. These studies include the dynamics of the Earth as a whole and of its component parts, volc-anism and rock formation, the hydrological cycle including snow and ice, all aspects of the oceans, the atmosphere, the ionosphere, the magnetosphere, and solar–terrestrial relations.

International Union of Geological Sciences (IUGS)
The IUGS was founded in 1961 in response to a need to co-ordinate geo-scientific international research programmes on a continuing basis in between the International Geological Congresses which are traditionally held every four years. IUGS aims to encourage the study of geoscientific problems, particularly those of worldwide significance, facilitate inter-national and interdisciplinary co-operation in geology and its related sciences, provide continuity to such co-operation, and support and provide scientific sponsorship to the quadrennial International Geological Congresses.

International Union of Microbiological Sciences (IUMS)
Founded in 1927 as the International Society of Microbiology, the IUMS became the International Association of Microbiological Sciences, affiliated to IUBS as a Division in 1967, acquired independence in 1980, and became a Union member of ICSU in 1982.

The objectives of the Union are to: promote the study of micro-biological sciences internationally; initiate, facilitate and co-ordinate research and other scientific activities which involve international co-operation; ensure the discussion and dissemination of the results of international co-operative research; promote the organization of inter-national conferences, symposia and meetings and assist in the publi-cation of their reports; and represent microbiological sciences in ICSU and maintain contact with other international organizations.

International Union of Pure and Applied Chemistry (IUPAC)

Established in 1919, IUPAC is the international body which represents chemistry amongst the other disciplines of science. The objectives of IUPAC are: to promote continuing co-operation amongst the chemists of its member countries; to study topics of international importance to pure and applied chemistry which need regulation, standardization or codification; to co-operate with other international organizations which deal with topics of a chemical nature; and to contribute to the advancement of pure and applied chemistry in all its aspects.

IUPAC is recognized throughout the world as the international authority on chemical nomenclature, terminology, symbols, units, atomic weights, and related topics. Its reports and recommendations on these matters are generally accepted as authoritative and definitive.

International Union of Pure and Applied Physics (IUPAP)

Established in 1922, the first General Assembly was held in 1923 in Brussels. The aims of the Union are: to stimulate and promote inter-national co-operation in physics; to sponsor international conferences; to foster the preparation and the publication of abstracts of papers and tables of physical constants; to promote international agreements on the use of symbols, units, nomenclature and standards; to foster free circ-ulation of scientists; and to encourage research and education in physics.

International Union of the History and Philosophy of Science (IUHPS)

Created in 1956 by the federation of the International Union of History of Science, which was founded in 1947, and the International Union of Philosophy of Science, founded in 1949. The two federating bodies became two Divisions of the Union: the Division of History of Science (DHS) and the Division of Logic, Methodology and Philosophy of Science (DLMPS).

The aims of IUHPS are to establish and promote international contacts among historians and philosophers of science and scientists who are interested in the history and foundational problems of their discipline; to collect documents useful for the development of history and philosophy of science; to encourage and sustain research and study of important problems in these fields; and to organize and support international conferences, symposia and other forms of scientific exchange.

International Ursigram and World Days Service see Federation of Astronomical and Geophysical Data Analysis Services

International Years of the Quiet Sun (IQSY)

An international co-operative programme of ICSU for the study of the Earth's environment based on observations made during 1964 and 1965, when solar activity (the number of sunspots) was at a minimum. IQSY covered meteorology, geomagnetism, aurora, airglow, ionosphere, solar activity, cosmic rays, aeronomy and space research.

interstadial

An interstadial is a "short" interval of geologic time (of the order of 1,000 years) within a glacial stage, during which the general climate remained cold but was not actually glacial. Interstadial periods separated successive maxima of glacierization within a single glacial stage.

intertropical convergence zone (ITCZ)

The zone of convergence between the airstreams from the Northern and Southern Hemispheres. It oscillates in position with the thermal Equator, reaching its maximum northward extent in July and maximum southward extent in January. Ascent of air causes low atmospheric pressure, deep convective clouds and heavy precipitation. Over the ocean areas it marks the meeting point of the trade winds, but over the continents it becomes affected by the monsoonal circulation and has somewhat different properties.

Inter-Union Commission on the Application of Science to Agriculture, Forestry and Aquaculture (CASAFA)

This Commission was established in 1978 at the 17th General Assembly of ICSU, to promote co-operation in research between institutions in developed and developing countries towards the resolution of problems of food production and processing in developing countries.

Inter-Union Commission on the Lithosphere (ICL)

In September 1980, the 18th General Assembly of ICSU approved the establishment of the Inter-Union Commission on the Lithosphere, with the objective of undertaking an international programme of interdisciplinary research for an improved understanding of the Earth, especially those aspects on which human society depends for its wellbeing. The programme is primarily concerned with the current state, origin, evolution, and dynamics of the lithosphere, with special attention to the continents and their margins. One special goal is to strengthen the Earth sciences and to make more effective their application in the developing countries.

inversion

A departure from the usual decrease or increase with height of an atmospheric property, most commonly referring to a temperature inversion when temperatures increase with height.

IOC see **Intergovernmental Oceanographic Commission**

IODE see **International Oceanographic Data Exchange**

ionosphere

The layer of the atmosphere above the stratopause in which free ions and electrons occur as a result of ionization of gas molecules by solar ultraviolet and X-radiation. The ionosphere consists of a series of constantly changing layers of heavily ionized molecules, termed the D-layer, E-layer, F-1 layer, and F-2 layer, which are superimposed upon the other non-ionized atmospheric regions.

IPCC see **Intergovernmental Panel on Climate Change**

IPCC First Assessment Report: WG I (Science)

The First Assessment Report of WG I (Science) of IPCC states:
We are certain of the following:
There is a natural greenhouse effect which already keeps the Earth warmer than it would otherwise be.

Emissions resulting from human activities are substantially increasing the atmospheric concentrations of the greenhouse gases, carbon dioxide, methane, chlorofluorocarbons (CFCs) and nitrous oxide. These increases will enhance the greenhouse effect, resulting on average in an additional warming of the Earth's surface. The main greenhouse gas, water vapour, will increase in response to global warming and further enhance it.
We calculate with confidence that:
Some gases are potentially more effective than others at changing climate, and their relative effectiveness can be estimated. Carbon dioxide has been responsible for over half of the enhanced greenhouse effect in the past, and is likely to remain so in the future.

Atmospheric concentrations of the long-lived gases (carbon dioxide, nitrous oxide and the CFCs) adjust only slowly to changes of emissions. Continued emissions of these gases at present rates would commit us to increased concentrations for centuries ahead. The longer emissions continue to increase at present-day rates, the greater reductions would have to be for concentrations to stabilize at a given level.

For the four scenarios of future emissions which IPCC has developed as assumptions (ranging from one where few or no steps are taken to limit emissions viz., Scenario A or Business-as-Usual Scenario, through others with increasing levels of controls, respectively called Scenarios B, C and D), there will be a doubling of equivalent carbon dioxide concentrations from pre-industrial levels by about 2025, 2040 and 2050 in Scenarios A, B, and C respectively.

Stabilization of equivalent carbon dioxide concentrations at about twice the pre-industrial level would occur under Scenario D towards the end of the next century. Immediate reductions of over 60% in the net (sources minus sinks) emissions from human activities of long-lived gases would achieve stabilization of concentration at today's levels; methane concentrations would be stabilized with a 15–20% reduction.

The human-caused emissions of carbon dioxide are much smaller than the natural exchange rates of carbon dioxide between the atmosphere and the oceans, and between the atmosphere and the terrestrial system. The natural exchange rates were, however, in close balance before human-induced emissions began; these steady anthropogenic emissions into the atmosphere represent a significant disturbance of the natural carbon cycle.

Based on current model results, we predict:

An average rate of increase of global mean temperature during the next century of about 0.3°C per decade (with an uncertainty range of 0.2–0.5°C per decade) assuming the IPCC Scenario A (Business as Usual) emissions of greenhouse gases; this is a more rapid increase than seen over the past 10,000 years. This will result in a likely increase in the global mean temperature of about 1°C above the present value by 2025 (about 2°C above that in the pre-industrial period), and 3°C above today's value before the end of the next century (about 4°C above pre-industrial). The rise will not be steady, because of other factors.

Under the other IPCC emissions scenarios which assume progressively increasing levels of controls, rates of increase in global mean temperature of about 0.2°C per decade (Scenario B), just above 0.1°C per decade (Scenario C), and about 0.1°C per decade (Scenario D). The rise will not be steady, because of other factors.

Land surfaces warm more rapidly than the oceans, and higher northern latitudes warm more than the global mean in winter.

The oceans act as a heat sink and thus delay the full effect of a greenhouse warming. Therefore, we would be committed to a further temperature rise which would progressively become apparent in the ensuing decades and centuries. Models predict that as greenhouse gases increase, the realized temperature rise at any given time is between 50% and 80% of the committed temperature rise.

With regard to uncertainties, we note that:

There are many uncertainties in our predictions, particularly with regard to the timing, magnitude, and regional patterns of climate change, especially changes in precipitation. These are due to our incomplete understanding of sources and sinks of greenhouse gases and the responses of clouds, oceans and polar ice sheets to a change of the radiative forcing caused by increasing greenhouse gas concentrations.

These processes are already partially understood, and we are confident that the uncertainties can be reduced by further research. However, the complexity of the system means that we cannot rule out surprises.

Our judgement is that:

Global mean surface-air temperature has increased by 0.3°C to 0.6°C over the past 100 years, with the five warmest years in terms of the global average being in the 1980s. Over the same period, global sea level increased by 10–20cm. These increases have not been smooth in time, nor uniform over the globe.

The size of the warming over the past century is broadly consistent with the prediction by climate models, but is also of the same magnitude as natural climate variability. If the sole cause of the observed warming were the human-made greenhouse effect, then the implied climate sensitivity would be near the lower end of the range inferred from models. Thus, the observed increase could be largely due to this natural variability; alternatively this variability and other human factors could have offset a still larger human-induced greenhouse warming. The unequivocal detection of the enhanced greenhouse effect from observations is not likely for a decade or more.

Measurements from ice cores going back 160,000 years show that the Earth's temperature closely paralleled the amount of carbon dioxide and methane in the atmosphere. Although we do not know the details of cause and effect, calculations indicate that changes in these greenhouse gases were part, but not all, of the reasons for the large (5–7°C) global temperature swings between ice ages and interglacial periods.

Natural sources and sinks of greenhouse gases are sensitive to a change in climate. Although many of the response (feedback) processes are poorly understood, it appears that, as climate warms, these feedbacks will lead to an overall increase, rather than a decrease, in natural greenhouse gas abundances. For this reason, climate change is likely to be greater than the estimates given above.

IPCC First Assessment Report: WG II (Impacts)

The First Assessment Report of WG II (Impacts) of IPCC states:
Any predicted effects of climate change must be viewed in the context of our present dynamic and changing world. Large-scale natural events such as El Niño can cause significant impacts on agriculture and human settlement. The predicted population explosion will produce severe impacts on land-use and on the demands for energy, fresh water, food, and housing, which will vary from region to region according to national incomes and rates of development. In many cases, the impacts will be felt most severely in regions already under stress, mainly the developing countries. Human-induced climate change due to continued uncontrolled emissions will accentuate these impacts. For instance, climate change, pollution, and ultraviolet-B radiation from ozone depletion can interact, reinforcing their damaging effects on materials and organisms. Increases in atmospheric concentrations of greenhouse gases may lead to irreversible change in the climate which could be detectable by the end of this century.

Comprehensive estimates of the physical and biological effects of climate change are difficult at the regional level. Confidence in regional estimates of critical climatic factors is low. This is particularly true of precipitation and soil moisture, where there is considerable disagreement between various general-circulation model and palaeo-analogue results. Moreover, there are several scientific uncertainties regarding the relationship between climate change and biological effects and between these effects and socio-economic consequences.

This impact study part of the overview does not attempt to anticipate any adaptation, technological innovation or any other measures to diminish the adverse effects of climate change that will take place in the same timeframe. This is especially important for heavily managed sectors, e.g. agriculture, forestry and public health.

The issues of timing and rates of change need to be considered; in particular, there will be lags between: (a) emissions of greenhouse gases and doubling of concentrations, (b) doubling of greenhouse gas concentrations and change in climate, (c) changes in climate and resultant physical and biological effects, and (d) changes in physical and ecological effects and resultant socio-economic (including ecological) consequences. The shorter the lags, the less the ability to cope and the greater the socio-economic impacts.

There is uncertainty related to these time lags. The changes will not be steady, and surprises cannot be ruled out. The severity of the impacts will depend to a large degree on the rate of climate change.

IPCC First Assessment Report: WG III (Response Strategies)

The First Assessment Report of WG III (Response Strategies) of IPCC states:
The consideration of climate change response strategies presents formidable difficulties for policy-makers. The information available to make sound policy analyses is inadequate because of (a) uncertainty with respect to how effective specific response options or groups of options would be in actually averting potential climate change; and (b) uncertainty with respect to the costs, effects on economic growth, and other economic and social implications of specific response options or groups of options.

The IPCC recommends a programme for the development and implementation of global, comprehensive and phased action for the resolution of the global warming problem under a flexible and progressive approach.

A major dilemma of the issue of climate change due to increasing emission of greenhouse gases in the atmosphere is that actions may be required well before many of the specific issues raised can be analyzed more thoroughly by further research.

The CFCs are being phased out to protect the stratospheric ozone layer. This action will also effectively slow down the rate of increase of radiative forcing of greenhouse gases in the atmosphere. Every effort should be made to find replacements that have little or no greenhouse-warming potential or ozone-depletion potential rather than the HCFCs and HFCs that are now being considered.

The single largest anthropogenic source of radiative forcing is energy production and use. The energy sector accounts for an estimated 46% (with an uncertainty rate of 38–54%) of the enhanced radiative forcing resulting from human activities.

It is noted that emissions due to fossil fuel combustion amounts to about 70–90% of the total anthropogenic emissions of carbon dioxide into the atmosphere, whereas the remaining 10–30% is due to human use of terrestrial ecosystems. A major decrease of the rate of deforestation as well as an increase in afforestation would contribute significantly to slowing the rate of increase of carbon dioxide concentrations in the atmosphere, but it would be well below that required to stop it. This underlines that when forestry measures have been introduced, other measures to limit or reduce greenhouse emissions should not be neglected.

IPCC Information Exchange Seminar Series

The IPCC Special Committee on the Participation of Developing Countries recommended steps to encourage the full participation of developing countries in the work of IPCC. One step was the dissemination of information on climate change issues by means of information exchange seminars.

Teams of 2 or 3 speakers familiar with the IPCC First Assessment Report are supported to present a 1–3 day seminar in each interested developing country. The IPCC Secretariat works closely with local organizers to encourage the participation of as many ministers (e.g. environment, energy, foreign affairs, agriculture, transportation, development and planning, and those responsible for the meteorological/hydrological services) and their senior advisers as possible. Sessions may also be held for other interested audiences, such as the members of the academia, the press and the industrial and environmental organizations. By December 1991, seminars had been held in several countries.

IPCC: Fifth Session – March 1991 and Work Programme for 1991

At the meeting of the fifth session of IPCC held in Geneva on 13–15 March 1991, it was decided to undertake work in the following six areas:
Task 1: Assessment on national net greenhouse gas emissions
The Panel noted that this topic divides naturally into three subsections, though with significant overlap. All three subsections should be under the guidance of a Steering Group with a balanced representation of governments; this Steering Group operating under the aegis of Working Group I should ensure the participation of Working Groups II and III as appropriate in this task.
Subsection 1: Sources and sinks of greenhouse gases.
The objective here is to improve the quantitative assessment of all sources and sinks (anthropogenic and natural, atmospheric, terrestrial and oceanic) of all the greenhouse gases (carbon dioxide, methane, nitrous oxide, halocarbons), tropospheric ozone precursors (carbon monoxide, nitrogen oxides, volatile organic compounds), and sulphur gases.
Subsection 2: Global Warming Potentials (GWPs)
The objective here is to develop further the concept of the global

warming potential, as an index of CO_2-equivalence, for all greenhouse gases and their precursors; and to compile an updated table of GWPs, with associated uncertainties.

Subsection 3: Emissions scenarios

The objective here is to update existing global scenarios of greenhouse gas and precursor emissions in the light of recent developments and adopted policies. This would enable a range of climate change predictions to be carried out.

Task 2: Predictions of the regional distributions of climate change and associated impact studies, including model validation studies.

The Panel noted the widespread and urgent requirement that existed for information on probable future climate changes at the regional and local levels and for corresponding assessments of regional and local ecological and socio-economic impacts. There was also a need to integrate regional-scale analyses into global impacts for sectors such as agriculture and forests.

The Panel recognized that some of the spatial and temporal variability of climate change may well be of a stochastic nature and therefore not predictable. It is obviously important to try to understand to what extent and in which way this may influence current ability to predict the regional characteristics of an anthropogenic climate change. The Panel emphasized that this aspect of the problem be studied and that the implications for predictions of future regional impacts of climate change be analyzed.

With a view to preparing a comprehensive revised assessment in 1994–95, the Panel agreed that the linkage between its Working Group I initiatives on regional climate change prediction and its Working Group II activities on regional impact assessment needed to be strengthened and made more effective than was possible in the compressed timescale of the First Assessment Report.

Task 3: Issues related to energy and industry

The purpose of the task is to fill significant gaps in the analysis achieved to date on the energy and industry sector responses to limit climate change; and to begin new areas of analysis and suggest areas of research which were too difficult to address in the first phase of the IPCC.

Task 4: Forestry-related issues

The Panel noted that human actions in the areas of forests and agriculture as the main types of land used contributed to net greenhouse gas (GHG) emissions on the one hand and were affected by climate change on the other. Both sectors were closely linked and of great importance for the whole ecosystem as well as for many human needs and activities. Therefore an integrated approach was necessary. Further, noting that land was a scarce resource, methodologies to assess and evaluate possible strategies and measures for its sustainable use needed to be developed.

Task 5: Vulnerability to sea-level rise

The Panel recalled the conclusions of its First Assessment Report, in particular (a) the commitment to continuing sea-level rise even after stabilizing the greenhouse forcing, and (b) the many national requests for assistance for assessing their vulnerability to sea-level rise including the identification of appropriate adaptive measures and the development of comprehensive national plans.

The Panel further recalled that a primary recommendation made by the Coastal Management Subgroup (CZMS) of WG III in the IPCC First Assessment Report was for coastal countries to formulate, by the year 2000, coastal management plans that incorporate response measures to reduce vulnerability to sea-level rise and address other immediate coastal resource management concerns. Since then, the interest in continuing the work on responses to sea-level rise had been strongly expressed by both developed and developing countries in informal discussions at various international meetings. These discussions had led the Panel to conclude that a new international effort, under the auspices of the IPCC and in co-operation with the UNEP Regional Seas Programme, should begin to assess the vulnerability of developing and developed countries to sea-level rise. The initial phase of the assessment, which would include, *inter alia*, methodology tested by a number of case-studies, should be completed as required by the INC process.

Task 6: Emissions scenarios

The Panel recalled that WG III had developed a set of emissions scenarios, that were subsequently used by the WG I for assessments of future climate change. These emissions scenarios were also used extensively by individual countries to support their national policy evaluations. The Panel agreed that recent developments warranted an update of the emissions scenarios.

The main goal in this task was the accommodation of relevant developments, both in science and policy, in the emissions scenarios. The results would be used for updates of the work of Working Group I. They might also be used as bases for transient GCM-calculations. In the long run, they would provide further information against which progress in controlling global emissions over time could be evaluated.

IPCC: **Sixth Session, October 1991**

The Intergovernmental Panel on Climate Change (IPCC) held its Sixth Session at Geneva from 29–31 October 1991. A key issue discussed at the Sixth Session was the relationship between the IPCC and the Inter-governmental Negotiating Committee for a Framework Convention on Climate Change (INC). The Panel agreed that its three Working Groups should focus closely on the problem of sinks of greenhouse gases as a contribution to the negotiating process for a climate convention. Other issues discussed as contributions to INC were the vulnerability to sea-level rise and extreme climatic occurrences. Special attention will be given to these issues which are of significance to INC. IPCC will also produce a supplement to its 1990 First Assessment Report, with additional input relevant to INC, early in 1992.

The meeting agreed that IPCC will continue to focus on the following tasks:
* assessment of national emissions of greenhouse gases, greenhouse gas warming potentials, and emissions scenarios;
* refining the ability to predict regional patterns of climate change;
* issues related to energy and industry;
* the contribution of agriculture and forestry to greenhouse gas emissions;
* vulnerability to sea-level rise.

IQSY see **International Years of the Quiet Sun**

isallobar
A line joining points of equal pressure-change during a given time interval.

isanomaly
A line along which the anomaly of a meteorological element has the same value.

ISLSCP
The International Satellite Land Surface Climatology Project.

ISCCP see **International Satellite Cloud Climatology Project**

isobar
A line on a weather map connecting points of equal atmospheric pressure.

isohyet
A line on a map connecting points receiving equal precipitation during a stated period.

isopleth
A line on a map connecting points at which given variables have the same numerical value (e.g. topographic contour lines).

isostacy
A condition of theoretical balance for all large portions of the Earth's crust, which assumes that they are floating on an underlying more dense medium. As a result of erosion or deposition, this balance is put out of equilibrium and has to be compensated for by movements of the Earth's crust. Areas of deposition sink, whereas areas of erosion rise.

isotherm
A line on a map connecting points having the same temperatures at a

particular time or during a stated period.

isotach
A line passing through points having the same value of windspeed.

isotach analysis
An analysis of the distribution of windspeeds on a reference surface by means of isotachs.

ISSS see **International Society of Soil Science**

ISY see **International Space Year**

ITCZ see **intertropical convergence zone**

ITTA see **International Tropical Timber Agreement**

ITTO see **International Tropical Timber Organization**

IUB see **International Union of Biochemistry**

IUBS see **International Union of Biological Sciences**

IUCN see **International Union for Conservation of Nature and Natural Resources**

IUFRO see **International Union of Forestry Research Organizations**

IUGG see **International Union of Geodesy and Geophysics**

IUGS see **International Union of Geological Sciences**

IUHPS see **International Union of the History and Philosophy of Science**

IUMS see **International Union of Microbiological Sciences**

IUPAC see **International Union of Pure and Applied Chemistry**

IUPAP see **International Union of Pure and Applied Physics**

IUPESM see **International Union for Physical and Engineering Sciences in Medicine**

J

jet streams

Cores of fast-moving air, with speeds of 200–300km per hour, which occur near the tropopause in temperate latitudes. They are a few thousand metres in depth and some tens of kilometres wide and move in an irregularly wavy pattern, from west to east in both hemispheres.

JGOFS see below

Joint Global Ocean Flux Study (JGOFS)

The Joint Global Ocean Flux Study (JGOFS), a core project of IGBP, is an internationally co-ordinated programme, organized by ICSU's Scientific Committee on Oceanic Research, to be carried out during the period 1990–2000. The objective of JGOFS is to understand the global biochemical and geochemical cycling of carbon and other biologically important chemical elements, which play a major rôle in the global environment. An essential component of JGOFS is a suite of basic physical and chemical measurements, to be carried out along representative oceanic sections, in co-operation with the WCRP World Ocean Circulation Experiment. Satellite remote sensing of ocean colour will also be undertaken, to characterize the marine primary biomass production and the fixation of CO_2 by the ocean.

Joint Group of Experts on the Scientific Aspects of Marine Pollution (GESAMP)

GESAMP is a multidisciplinary body of independent experts which provides advice to the sponsoring organizations – IMO, FAO, UNESCO, WMO, WHO, IAEA, UN, UNEP – at their request, on pollution and other problems facing marine and coastal environments. For the purpose of its earliest deliberations, GESAMP defined marine pollution as the introduction by human activities, directly or indirectly, of substances or energy into the marine environment (including estuaries) which result in such deleterious effects as to cause harm to living resources, hazards to human health, hindrance to marine activities (including fishing), impairment of the quality of sea water, and the reduction of amenities.

In recent years, as GESAMP has begun to broaden its concerns to include protection and management of marine and coastal environments, it has adopted the concept of sustainable development. This concept implies that the present use of the environment and its resources shall not prejudice the use and enjoyment of that environment and its resources by future generations.

Most of the work of GESAMP is carried out by Working Groups. During the first two decades of its existence, 32 GESAMP Working Groups have been formed including:

* Working Group 14 (interchange of pollutants between the atmosphere and the oceans);
* Working Group 24 (integrated global ocean monitoring);
* Working Group 26 (state of the marine environment);
* Working Group 28 (scientifically based strategies for marine environmental protection and management);
* Working Group 31 (environmental impacts of coastal aquaculture);
* Working Group 32 (global change and the air/sea exchange of chemicals).

Joint Scientific and Technical Committee (JSTC)

The functions of the Joint Scientific and Technical Committee (JSTC) are to formulate the overall concept and scope of the Global Climate Observing System (GCOS), and to provide scientific and technical guidance to sponsoring and participating organizations and agencies for the planning and further development of the GCOS. Specifically, the JSTC will be called upon:

* to identify observational requirements, define design objectives and recommend co-ordinated actions by sponsoring and participating organizations and agencies, in order to optimize the system's performance and coherence, taking cognizance of the responsibilities, working arrangements and recommendations of established scientific and technical bodies of such organizations and agencies;
* to review and assess the development and implementation of the components of the GCOS, and report to the sponsoring organizations, and to the participating agencies as required;
* to facilitate the exchanges of information among sponsoring and participating organizations and agencies and, in general, make the objectives, resource requirements and capabilities of GCOS known to relevant national and international bodies.

The JSTC shall be appointed jointly by the Executive Heads of WMO, IOC and ICSU, after consultation with other relevant international organizations and participating agencies.

Joint Scientific Committee of the World Climate Research Programme (WCRP)

The Joint Scientific Committee (JSC) of the World Climate Research Programme (WCRP) was established by WMO and ICSU in 1980 to oversee the work of the WCRP. It meets once a year.

JSC see above

JSTC see **Joint Scientific and Technical Committee**

K

katabatic wind

A downslope wind caused by greater air density along the slope than at some distance horizontally from it. The wind is associated with surface cooling of the slope.

Krakatoa

Volcanic islet between Java and Sumatra, which erupted very violently on 20 May 1883, emitting very large amounts of particulate matter and gases into the atmosphere.

L

La Niña see El Niño

lapse rate

The rate of decrease of an atmospheric variable (such as temperature or moisture) with height.

latent heat

The energy required to change water from the liquid to gaseous phase and vice versa. The energy transferred from the Earth's surface to the atmosphere through the evaporation and condensation processes is an important form of latent heat.

laterite

The red and highly weathered residual soil, characteristic of moist tropical and subtropical regions, which is rich in the oxides of iron and aluminium.

Law of the Atmosphere (LoA)

A proposed/possible "law" which would contain rules of access for the common resource of the global atmosphere. Like the Law of the Sea it could/would cut across a number of issues in a comprehensive way. In addition to climate change, a Law of the Atmosphere could accommodate other global and regional atmospheric issues. Most importantly, a Law of the Atmosphere could directly couple financial mechanisms to the use of the atmospheric common resource. Because the package could be comprehensive, it could in theory accommodate the wide range of equity concerns.

LDC see **less developed country**

leaching
The process by which nutrient chemicals or contaminants are dissolved and carried away by water, or are moved into a lower layer of soil.

lee waves
Waves in the airstream to the lee of hills and mountains.

less developed country (LDC)
"less developed country" (LDC) is a term used in statistical compilations of UN institutions to designate developing countries.

lightning
A luminous manifestation accompanying a sudden electrical discharge which takes place from or inside a cloud or, less often, from high structures on the ground or from mountains.

limitation strategies see **management options for responding to climate change**

limnology
The science that deals with the study of lakes and open reservoirs, including hydrological phenomena, emphasizing the analysis of the environment.

lithosphere
The upper (oceanic and continental) layer of the solid Earth, comprising all crustal rocks and the brittle part of the uppermost mantle. Its thickness is variable, from 1–2km at mid-oceanic ridge crests, but generally increases from 60km near the ridge to 120–140km beneath older oceanic crust.

Little Ice Age
A period between about 1550 and 1860 during which the climate of the middle latitudes became generally harsher and there was an expansion of glaciers. The effects have been recorded in the Alps, Norway and Iceland, where farm land and buildings were destroyed. There were times of special severity during this period, e.g. the early 1600s when glaciers were very active in the Chamonix valley, in the French Alps.

LoA see **Law of the Atmosphere**

logged-over forest
Virgin or natural forest which has been generally systematically exploited for the extraction of timber.

London Ministerial Conference on Ozone (1990)

The Ministerial Conference held in London in June 1990 on revising the Montreal Protocol on the protection of the ozone layer agreed (among other things) to eliminate the production and use of CFCs by the year 2000, with a reduction of 85% by 1997. A group of 13 countries also made a separate declaration to phase out CFCs as soon as possible and by 1997 at the latest. There was also an agreement to set up a US$240 million fund to help Third World countries obtain technology from the industrialized world to develop alternatives to CFCs. This was the first time that a specific international fund has been actually set up for environmental purposes. The fund, initially to run for three years, will be administered by the World Bank together with the UNEP and UNDP.

long wave

An atmospheric "wave" in the major belt of westerlies which is characterized by large length and significant amplitude. The wavelength is typically longer than that of the rapidly moving individual cyclonic and anticyclonic disturbances of the lower troposphere.

longwave radiation

The radiation emitted in the spectral wavelength corresponding to the radiation emitted from the Earth and atmosphere. It is sometimes referred to as "terrestrial radiation" or "infrared radiation", although somewhat imprecisely.

low

An area of low atmospheric pressure with a closed anti-clockwise circulation in the Northern Hemisphere and closed clockwise circulation in the Southern Hemisphere; also called a depression in some areas of the world.

low index situation

The state of the westerly circulation when flow is weak and there is appreciable meridional transfer, as in a blocking situation.

M

MAB

The Man and the Biosphere Programme (UNESCO).

macroclimate

The general large-scale climate of a large area or country, as distinguished from the mesoclimate and microclimate.

management options for responding to climate change

Strategies for responding to a climate change as a result of increasing amounts of greenhouse gases in the atmosphere fall into two categories. Limitation strategies slow or reverse the growth of greenhouse gas concentrations in the atmosphere, whereas adaptation strategies adjust the physical environment or our ways of using it to reduce the consequences of a changing climate. A prudent response should consider both approaches.

Adaptation will involve expenditure of very large financial resources over long periods of time with extended advance planning. Measures to adapt to climate change will occur on a variety of scales and with widely varying costs. Some environmental modification measures, such as changes in coastal defences and freshwater supply systems, require infrastructure investments beginning decades in advance of anticipated climate impacts. On the other hand, many adjustments will consist of behavioural changes at the individual level occurring in immediate response to perceived warming, with little advanced planning.

Various limitation options exist for achieving carbon dioxide reductions, in spite of continued population growth, and in a manner consistent with continued economic expansion. These include a reversal of the current deforestation trend, efficiency advances in generation and transmission of energy, shifting the fossil-fuel use mix from high to low carbon-dioxide-emitting fuels, disposal of carbon dioxide in the deep ocean, and replacing fossil-fuel combustion with alternative energy sources such as solar, wind, hydroelectric, nuclear, tidal and ocean thermal conversion.

However, even if limitation and adaptation strategies are found to give cost-effective management solutions to the climate change problem on a global scale, the actions to achieve both strategies will have to be taken at national and local levels and in many cases in accordance with internationally agreed guidelines. Moreover, every nation will differ in the economic and social costs which it has to devote to limitation and adaptation strategies. These costs will depend on a number of factors: the degree and type of industrialization and fossil-fuel use, and the gross domestic product per capita; the type and level of nitrogenous fertilizer used in agriculture; the geographical location of the country; and whether there is a coastline (including how much of it is low lying, built over, or used for agriculture or tourism).

Some countries believe that in the context of climate change they will be winners and some believe they will be losers with a problem not of their own making. At present, the information available on climate impacts is not sufficient to suggest clear winners and losers, but some countries could well perceive that they will be.

mangrove forest

Tropical or subtropical inundation forests growing in the tidal zone along shallow coasts and river courses.

Markov chain

A Markov process in which the time parameter is discontinuous.

Markov process

A stochastic process such that the conditional probability distribution for the state at any future instant, given the present state, is unaffected by any additional knowledge of the past history of the system.

market forces and greenhouse gases

A preference for adaptation as a strategy in the greenhouse gas/climate change dilemma is often expressed by those who consider that free market forces will take care of the problems of adaptation, as and when they may occur, whereas this could not happen with a pre-planned limitation strategy. The difficulty here is that the long timeframes of climate and ecological processes do not fit into the much shorter timeframes of economics, where events 10 years off can have their values discounted by 65% and those 40 years off by 98%. It is considered that the discounting problem will ensure that the rôle of free-market mechanisms will be at best marginal, and cannot take the place of government actions at both national and international levels.

Marine Environmental Data Referral System (MEDI)

MEDI is a project of IOC in which data centres, international organizations and marine scientific institutions worldwide have provided input concerning the availability, location and characteristics of their marine environmental data.

MARS

A Netherlands-sponsored programme concerned with monitoring agro-ecological resources by means of remote sensing and simulation.

Mauna Loa carbon dioxide record

The record of measurements of atmospheric carbon dioxide concentrations taken at the Mauna Loa Observatory, Mauna Loa, Hawaii, beginning in March 1958 and continuing to the present. The Mauna Loa record is the longest reliable daily record of atmospheric carbon dioxide measurements in the world.

Maunder minimum see sunspot minima

maximum (minimum) temperature

Highest (lowest) temperature attained during a given time interval.

mean annual maximum (minimum) temperature

Mean of the daily maximum (minimum) temperatures observed in a specific year or over a specified period of years.

mean annual range of temperature
Difference between the mean temperatures of the warmest and coldest months of the year.

mean daily maximum (minimum) temperature for a month
Mean of the daily maximum (minimum) temperatures observed during a given calendar month, either in a specified year or over a specified period of years.

mean daily temperature
Mean of the temperatures observed at 24 equidistant times in the course of a continuous interval of 24 hours; or a combination of temperatures observed at longer intervals, so arranged as to depart as little as possible from the mean defined above.

mean monthly maximum (minimum) temperature
Mean of the monthly maximum (minimum) temperatures observed during a given calendar month over a specified period of years.

mean sea level
The average height of the sea surface, based upon hourly observation of the tide height on the open coast or in adjacent waters that have free access to the sea. In the United States, it is defined as the average height of the sea surface for all stages of the tide over a nineteen-year period. Mean sea level, commonly abbreviated as msl and referred to simply as "sea level", serves as the reference surface for all altitudes in upper atmospheric studies.

MDD
METEOSAT Data Dissemination. A system used, among others, to encypher METEOSAT data, ending a free availability of data from meteorological satellites.

MECCA see Model Evaluation Consortium for Climate Assessment

MEDI see Marine Environmental Data Referral System

medical climatology
The branch of climatology which studies the influence of climate on the health and disease of human beings.

mega
A prefix meaning million.

mercury barometer
A barometer in which the atmospheric pressure is balanced against the

pressure exerted by a column of mercury.

meridional flow

In meteorology a term meaning longitudinal or along a meridian such as northerly or southerly; opposed to zonal which is westerly or easterly. It is also an atmospheric circulation in a vertical plane orientated along a meridian. It consists, therefore, of the vertical and the meridional (north or south) components of motion only.

mesoclimate

The climate of small areas of the Earth's surface which may not be representative of the general climate of the district.

mesosphere

The portion of Earth's atmosphere from about 50km to 80km in which the temperature, with increasing height, at first rises to a maximum of around 280°K and then decreases to around 180°K or less, depending on the latitude and the season. Its lower boundary is the stratopause and the upper boundary is the mesopause.

metadata

Station history files of climate data which are on a common computer-based format for those sites.

meteoric dust

Atmospheric dust originating from meteors.

methane

Methane is greenhouse gas produced by anaerobic decomposition of vegetal materials in wetlands and rice paddies, and also in the stomach of cattle, so that emissions are directly related to the development of agriculture and animal husbandry. It also occurs as leakage from natural gas pipelines. Human activities such as rice cultivation, domestic ruminant rearing, biomass burning, coal mining, and natural gas venting have increased the input of methane into the atmosphere which, combined with an possible decrease in the concentration of tropospheric OH, yields the observed rise in global methane.

methane concentration

Current atmospheric methane concentration, at 1.72ppmv is now more than double the pre-industrial (1750–1800) value of about 0.8ppmv, and is increasing at a rate of about 0.015ppmv (0.9%) per year. The major sink for methane, reaction with hydroxyl (OH) radicals in the troposphere, results in a relatively short atmospheric lifetime of about 10 years.

methane from rice paddies

At the present level of technology, yields are lower in upland rice (non-flooded) due to the lack of appropriate varieties and adequate agronomic experience. However, aerobic rice has the potential to reduce water pollution from nitrogen fertilizers and algaecides, and the release into the atmosphere of nitrous oxides (NO_x) and methane.

Most flooded rice is grown in areas originally occupied by swamps or water-logged soils. Methane emissions constitute a loss of carbon from the fields, and could be reduced by agricultural practices. It appears likely that methane and nitrous oxide emissions are linked in flooded rice. Thus, measures and techniques to limit methane would have a beneficial effect on nitrous oxides as well.

methane from ruminant digestion

Half of the exploited agricultural areas of the world are in pasture, and livestock living on this pasture is one of the major producers of methane. In particular ruminant animals (cattle, sheep, goats) which, it is estimated, generate approximately 80Gt of the gas per year, contribute the equivalent of 18% of the world's methane production. Many of these animals, and especially those extensively managed in developing countries of the world (about 570 million cattle and 140 million buffaloes), while consuming vast quantities of plant material, are very inefficient converters and yield very little animal product. At the same time they denude the ground-cover from large areas of the world, thereby reducing the capacity of rangelands to act as a sink for carbon dioxide.

In contrast, livestock production in the developed world has been intensified greatly over the past three decades but at the expense of a high input of fossil-fuel dependent products, such as, for example, mineral fertilizer, mechanized labour and chemicals. It is also important to realize that industrialized high-output animal production is not necessarily efficient in terms of animal bio-energetics or energy input:output ratios.

micro

A prefix meaning one millionth.

microclimate

The microclimate structure of the air space which extends from the surface of the Earth to a height where the effects of the immediate character of the underlying surface cannot be distinguished from the general local climate (mesoclimate or macroclimate).The microclimate can be subdivided into as many different classes as there are types of underlying surface. The most studied microclimatic types are: the "urban microclimate", affected by pavements, buildings, air pollution, dense habitation, etc.; the "vegetation microclimate", concerned with the complex nature of the air space occupied by vegetation, and its effects upon the vegetation; and the microclimate of the confined spaces, of houses, greenhouses, caves, etc.

Milankovitch solar radiation curve

A radiation curve which combines the effects of the precession of the equinoxes, the obliquity of the ecliptic, and the eccentricity of the Earth's orbit. Early in the twentieth century, a Yugoslavian mathematician and physicist M. Milankovitch calculated the composite solar radiation curve and used it to account for the variations of climate during the ice ages. He postulated that each period of radiation minimum caused an ice age. (See also **astronomical theory of climate change** and **Earth's orbital variations**)

milli

A prefix meaning one thousandth.

mist

The suspension in the air of microscopic water droplets or wet hygroscopic particles, reducing the visibility at the Earth's surface.

mitigation strategies see **management options for responding to climate change**

Model Evaluation Consortium for Climate Assessment (MECCA)

A new international consortium (MECCA), made up of the partnership of industry, government, and academic groups from the US, Japan, Italy and France, which is operating the first supercomputing laboratory in the United States devoted solely to modelling climate changes due to greenhouse gases. Researchers are using a new Cray Y-MP super-computer and the associated facilities at the National Center for Atmospheric Research in Boulder, Colorado, to improve our under-standing of the models being used to forecast global climate changes. MECCA acknowledges the concerns about greenhouse-gas- induced global climate change, and recognizes that policy-makers will act in response to those concerns. MECCA also supports research to quantify the uncertainty surrounding the climate models that will be the basis of that policy. MECCA will produce information which policy-makers can use to co-ordinate policy with scientific developments.

modelling

An investigative technique that uses a mathematical or physical representation of a system or theory that accounts for all or some of its known properties. Models are often used to test the effects of changes of system components on the overall performance of the system.

molecule
The smallest part of a substance that can exist separately and still retain its chemical properties and characteristic composition.

monitoring
The continuous or frequent standardized measurement and observation of the environment (see also **systematic observations**).

monsoon
A name for seasonal winds (derived from Arabic "mausim", a season). It was first applied to the winds over the Arabian Sea, which blow for six months from the northeast and for six months from the southwest, but it has been extended to similar wind systems in other parts of the world. The monsoons are strongest on the southern and eastern sides of Asia, but also occur on the coasts of tropical regions wherever the planetary circulation is not strong enough to inhibit them.

monthly climatological summary
Frequencies, means or totals of the observations made during the month, with additional relevant climatological statistics.

monthly maximum (minimum) temperature
Highest (lowest) temperature recorded in the course of a given calendar month in a specified year.

Montreal Protocol
An agreement, called the Montreal Protocol on Substances which Deplete the Ozone Layer, was signed, subject to ratification, by 24 countries on 16 September 1987. The Protocol is a complex legal document but the main points are: (a) by 1990, all parties to the Protocol will freeze CFC consumption to the same levels as 1986; (b) by 1994, all parties will reduce CFC consumption to 80% of the 1986 levels; and (c) by 1999, all parties will reduce consumption to 50% of the 1986 level.

If these consumption levels are achieved, it will mean that chlorine amounts will by 1999 increase by only 2% per year, compared with an increase of 5% per year if there were no Protocol. The Protocol will therefore limit the worsening of the ozone problem. Clearly, damage to the ozone layer will still continue, but there is now a clear signal to industry for action to develop alternatives and new technology.

There are various provisions for continuing scientific assessments and reviews of the Protocol timetable. It is important to note that in this regard the Protocol is consumption- rather than production-based. This means that the burden of compliance is shared more evenly between the producing nations and those which import all their CFCs.

See also **London Ministerial Conference on Ozone (1990)**

most-favoured-nation clause

A clause inserted into trade agreements by virtue of which any trade-policy advantages, especially customs advantages, offered to another signatory state must be granted to all other states as well. Favoured-nation clauses are inserted into many bilateral trade agreements, and such a clause is also binding for the members of the General Agreement on Tariffs and Trade. The intention is to protect the member states against discrimination.

Mount Agung

A volcano in Indonesia which erupted violently in March 1963, emitting large amounts of particulate matter and gases. Large quantities of sulphur dioxide, which reached a height of about 18km, spread over the world within about six months, and lasted for several years and produced twilight displays of colours.

Mount Pinatubo

This Philippine volcano started to erupt on 9 June 1991 and it is one of the major volcanic eruptions of the 20th century, if not the largest. Estimates from the Nimbus-7 satellite are that Pinatubo is emitting twice the sulphur dioxide of the 1982 El Chichon eruption. Debris from the volcano is expected to cool global temperatures for a two- to four-year period, and many scientists predict that 1992 and 1993 will be cooler, perhaps by several tenths of a degree Celsius.

Mount Tambora

A volcano in Indonesia which in 1815 produced the most violent eruption in historical times, reducing the height of the mountain from 4000m to 2800m, and emitting large quantities of gases and particles.

multivariate

Refers to a system in which there is simultaneous variation in more than one independent variable.

N

Nairobi Declaration on Climatic Change

At an International Conference on Global Warming and Climatic Change held in Nairobi, Kenya, from 2-4 May 1990, organized by the African Centre for Technology Studies and the Woods Hole Research Center, a "Call for Action" was included in a Declaration on Climatic Change. The "Call for Action" stated:

"The expected effects of climatic change on the African economies call for urgent action on the part of governments. In the past, humankind has relied on the practice of reacting and correcting. But climatic change requires new forms of practice which include anticipating and preventing the expected effects. So far there is no clear evidence on the issue of global warming. So it can be argued that the costs of taking preventive measures are too high to be justified without convincing evidence. But for Africa, such a position is untenable for three reasons. First, by the time the scientists generate convincing evidence, it might be too late to take any preventive measures. Secondly, the magnitude of the effects of climatic change would be so large as to require preventive measures to be taken at the earliest possible moment; that moment is now. Thirdly, most of the measures proposed for dealing with the problem are necessary anyway for shifting towards sustainable development. The current problems facing Africa need urgent attention. The realization that climatic change might worsen these problems calls for remedial measures to be introduced as a matter of priority."

nano
A prefix meaning one billionth (1×10^{-9}).

NASA
The National Aeronautics and Space Administration (USA).

national climate programmes
The purpose of officially recognized National Climate Programmes (programmes recommended by WMO through the World Climate Programme) is to build on the operational infrastructure and database within a country and to realize the material benefits from an interdisciplinary approach to a broad range of social and economic issues for which climate is an important element. The primary objectives of a National Climate Programme are:
* to strengthen the existing infrastructure and ensure the capability to apply existing climate information to the planning and management of national socio-economic and environmental programmes;
* to develop a capability to foresee significant climate variations, either natural or manmade, which can markedly affect national welfare.
Specific requirements of a National Climate Programme include:
* recognition and specification of high-priority national activities related to climate;
* development of a plan of action under each national objective;
* formal co-ordination of public activities (including the national, regional and local levels as necessary) to achieve the national objectives.
The development in each country of a well co-ordinated national climate programme should yield benefits that include: (a) a more effective climate information database for planning, management and research; (b) improved specification in engineering design and reduction of risk from infrastructure system failure as a result of climate extremes; (c)

improved tactical decision-making for climate-sensitive industries, particularly during prolonged anomalies such as drought; (d) better socio-economic planning and public response strategies for seasonal extremes such as prolonged heatwave, high fire danger, or severe storms; (e) a broader comprehension of the impact of human activity on the environment, both at the local level and in the context of potential global climate change; and (f) a base for developing long-term strategies to cope with a possibly changing climate, especially if the change is a direct consequence of human activity (e.g. the greenhouse effect).

national emission strategies

Strategies on emissions developed by individual nations. In this regard emission strategies adopted by individual political units within a nation could be more stringent than those at the national level.

natural selection

The process of "survival of the fittest", by which organisms that adapt to their environment survive and those that do not adapt to their environment disappear.

natural variability of climate

The natural variability of climate where natural is considered to be not influenced by any human or human-related activity. The real natural variability of the climate, especially in heavily populated areas is very difficult to determine, but in sparsely or unpopulated areas it is more easily determined. However, few if any parts of the world are completely immune from human influences.

NCAR

The National Center for Atmospheric Research at Boulder, Colorado, USA.

negative feedback

An interaction within a system that causes a reduction or dampening of the response of the system to a force.

negotiation

The process of seeking accommodation and agreement on measures and policies among two or more interests or agencies having initially conflicting positions, by a "voluntary" or "non-legal" approach.

net flux

The difference between the total incoming quantity of a material or energy and the total outgoing quantity. Net flux can be positive, negative, or zero.

net primary production

Over a specified period of time, the biomass which is incorporated into a plant community, measured by the net amount of atmospheric carbon sequestered by green plants. The mechanisms involved are absorption of carbon dioxide and the fixing of the carbon by photosynthesis.

net radiation

The difference between the downward (total) and the upward (terrestrial) radiation.

NGO see **non-governmental organization**

nitrous oxide

Nitrous oxide concentrations are now increasing by 0.3% per year, and present levels are about 5–10% greater than pre-industrial values. Although the natural cycle for nitrous oxide is poorly understood, the increased releases are believed to come primarily from ammonia-based fertilizers (both in chemical form and as natural wastes from domestic animals) and the burning of fossil fuels for energy.

NOAA

The National Oceanic and Atmospheric Administration (USA).

non-compliance measures

Legal or other measures which would or could be enforced (by a body to be determined) if a nation did not meet internationally agreed emission standards.

non-governmental organization (NGO)

An organization such as an environmental group, trade association, or research institute, usually with an international membership or involved in international activities, which (among other things) may be accorded official status by the UN, enabling it to attend certain UN meetings in an observer or consultant capacity, and to provide information and views to UN committees.

non-use of forests

The deliberate non-use of forests on the basis that appropriate compensation will be paid. The funding mechanism for such compensations would need to be established.

Noordwijk Declaration (1989)

In October 1989, ministerial-level participants from many countries throughout the world conferred in Noordwijk in the Netherlands and agreed on a declaration on "atmospheric pollution and climatic change". The declaration stated among other things that "The composition of the

Earth's atmosphere is being seriously altered at an unprecedented rate due to human activity. Based on our current understanding, society is being threatened by manmade changes to the global climate. . . . While there are still uncertainties regarding the magnitude, timing and regional effects of climatic change due to human activity, there is a growing consensus in the scientific community that significant climate change and instability are most likely over the next century. Predictions available today indicate potentially severe economic and social dislocations for future generations. Assuming these predictions, delay in action may endanger the future of the planet as we know it."

The Declaration also (among other things):

* Urges all countries and relevant organizations to increase their climate change research and monitoring activities and to provide for adequate databases on emissions. It also urges states to co-operate in, and provide increasing support for, international co-ordination of these activities, building on international programmes such as the World Climate Programme and the International Geosphere–Biosphere Programme, and on the present roles of the UNEP, WMO, ICSU, IEA, Unesco, IOC, and other competent international organizations and bodies. The enhancement and strengthening of operational aspects of their work should be examined.
* Recommends that more research should be carried out by 1992 into the sources and sinks of the greenhouse gases other than CO_2 and CFCs, such as methane (CH_4), nitrous oxide (N_2O) and tropospheric ozone (O_3), including further research on the effect of the ocean on the concentration of radiatively active gases in the atmosphere.

normalized series

In statistics, a series obtained by subtracting the mean from each term of a series and dividing the result by the standard deviation. The use of normalized series simplifies the comparison of data expressed in distinct units.

normals (climatic)

Period averages computed for a uniform period of 30 years.

North–South dialogue

A term used for endeavours involving the reconciliation of interests between the industrialized nations (the so-called "North") and the developing nations (the so-called "South").

nuclear fusion

A nuclear reaction in which light atomic nuclei are fused to form heavier nuclei accompanied by the release of energy as radiation.

nuclear power
Power derived from fission or fusion nuclear reactions. More conventionally, nuclear power is interpreted as the utilization of the fission reactions in a nuclear power reaction to produce steam for electrical power production, for ship propulsion or for process heat. Fission reactions involve the break-up of the nucleus of heavyweight atoms and yield energy release which is more than a millionfold greater than that obtained from chemical reactions involving the burning of a fuel. Successful control of the nuclear fission reactions provides for the utilization of this intensive source of energy, and with the availability of ample resources of uranium deposits, significantly cheaper fuel costs for electrical power generation are attainable.

nuclear reactor
A device for generating heat by the controlled fission of atoms of uranium-235 or by the fusion of light atoms.

nuclear winter
A widespread climatic cooling likely to occur as a result of nuclear warfare in which the amount of radiation penetrating to the Earth's surface would be reduced.

numerical weather forecast
An objective forecast, in which the future state of the atmosphere is determined by the numerical solution of the basic theoretical equations involved. Various atmospheric "models" are used for numerical forecasts, varying chiefly in the degree of complexity with which the vertical structure of the atmosphere is accounted for. The simplest and most restrictive is the barotropic model in which a single pressure surface, usually 500 hectopascals, is considered. (For this kind of model, the vorticity equation reduces to the statement that a fluid element conserves its absolute vorticity or spin throughout its history.) Alternatively, baroclinic models may be used which involve consideration of several surfaces, allowing vertical variations of winds and temperatures to be represented more realistically. Some "primitive-equation models" use ten or more surfaces.

O

OAU
The Organization of African Unity.

obliquity of the ecliptic
The angle between the plane of the ecliptic (or the plane of the Earth's

orbit) and the plane of the Earth's Equator; the "tilt" of the Earth. M. Milankovitch calculated that the obliquity of the ecliptic varies between 24.5° and 22° in the course of 40,000 years. This variation may be considered as a long-period climatic control. The present value of the obliquity is 23°27'.

Ocean-PC

Ocean-PC is a standard software package for oceanographic data processing and exchange developed by IOC. The purpose of the package is to:

* Identify marine science software applications to be shared, starting with packages well known in the International Oceanographic Data and Information Exchange (IODE) community, and extending the list by advertising in the IMS Newsletter, PC magazines, bulletin boards, etc. The software is to be non-commercial, although templates and macros for existing commercial packages may be submitted.

* Evaluate the applications identified with respect to the needs of the marine scientist, the improvements of global data exchange, and the functionality required for later inclusion into the Ocean-PC software by consideration of user-friendliness, quality of documentation, fulfilment of description, state of development, adaptability, support, security, etc.

* The software will be included into a library, distributed by IOC. The library will include tools, time and productivity packages, format conversion and exchange routines, etc., with particular focus on analytical capabilities, quality control and data visualization for marine science.

OECD

The Organization for Economic Co-operation and Development.

OHP see below

okta

A fraction equal to one-eighth of the celestial dome, used in the coding of cloud amounts in synoptic meteorological observations.

OPEC see **Organization of Petroleum Exporting Countries**

Operational Hydrology Programme (OHP) of WMO

The basic systems component of this WMO Programme concentrates on the basic organization and phased development of Hydrological Services. It includes development, comparison, standardization and improvement of instruments, and methods for the collection and archiving of water resources information (quantity and quality of both surface water and groundwater), and human resource development. Specific support to the transfer of technology is provided through the Hydrological Operational

Multi-purpose Subprogramme (HOMS).

The applications and environment component of the Programme brings together hydrological activities in support of water-resource development and management, including hydrological modelling and forecasting, and the provision of data for a range of projects, including those for environmental protection. It contributes to various meteorological and climatological programmes of WMO, such as the Tropical Cyclone Programme (TCP) and the World Climate Programme (WCP).

The third component of the Programme on water-related issues contributes to the international programmes of other bodies within the UN family and to those of intergovernmental organizations and NGOs through inter-agency co-ordination and collaboration in water-related activities, including regional projects associated with large international river basins.

Organisation for Economic Co-operation and Development (OECD)

Pursuant to Article 1 of the Convention signed in Paris on 14 December 1960, and which came into force on 30 September 1961, the Organisation for Economic Co-operation and Development (OECD) shall promote policies designed:
* to achieve the highest sustainable economic growth and employment and a rising standard of living in member countries, while maintaining financial stability, and thus to contribute to the development of the world economy;
* to contribute to sound economic expansion in member as well as non-member countries in the process of economic development; and
* to contribute to the expansion of world trade on a multilateral, non-discriminatory basis in accordance with international obligations.

The original member countries of the OECD are Austria, Belgium, Canada, Denmark, France, Germany, Greece, Iceland, Ireland, Italy, Luxembourg, the Netherlands, Norway, Portugal, Spain, Sweden, Switzerland, Turkey, the United Kingdom and the United States. The following countries became members subsequently: Japan, Finland, Australia and New Zealand. The Commission of the European Communities takes part in the work of the OECD.

Organization of Petroleum Exporting Countries (OPEC)

The Organization of Petroleum Exporting Countries (OPEC) was established in 1960. It is a group of Asian, African and Latin American countries which are major producers and exporters of crude petroleum.

Ottawa Meeting on Legal Issues of Climate Change

A meeting held in Ottawa (Canada) in February 1989 on the legal aspects of any protocol likely to be developed dealing with climate change.

over-exploitation

The unsustainable exploitation of a potentially renewable natural resource.

ozone

An ozone molecule consists of three atoms of oxygen. In contrast the normal oxygen in the atmosphere exists as a molecule with only two atoms of oxygen. Ozone is much more reactive than oxygen and is toxic to human beings and living matter. As a pollutant at ground level it is causing (among other things) some damage to forests. In the stratosphere, ozone functions as both a greenhouse gas and a filter for ultraviolet radiation. A fall in the total ozone concentration in the atmosphere, and consequent rise in the penetration of ultraviolet radiation, could cause deleterious effects, such as skin cancers.

The concern that resulted in governments signing the 1987 Montreal Protocol, with unusual speed, arises from the realization that the ozone layer in the stratosphere is being destroyed by manmade substances called chlorofluorocarbons (CFCs) and halons. These are used as refrigerating fluids and spray-can propellants because they are convenient and cheap. They are also very stable and they remain in the troposphere for many decades. But as they diffuse eventually into the stratosphere, the chlorine they contain is liberated and it catalyses the break-up of ozone. This has been known in theory since 1974, but the most dramatic pointer of ozone depletion – the Antarctic ozone "hole" – was first detected in 1985.

So far as the greenhouse gases are concerned, ozone absorbs infrared radiation, as does carbon dioxide, and it therefore contributes directly to the greenhouse effect. In addition, a decrease in total ozone resulting in increased ultraviolet radiation reaching the upper layers of the sea may cause the death of phytoplankton. If this happens, the marine biomass will be less able to absorb the carbon dioxide dissolved in the water, reducing the ocean's effectiveness as a carbon sink. The effect of this would be to leave more carbon dioxide free in the atmosphere.

ozone creation and destruction

In its natural state, atmospheric ozone occurs in a layer which has a maximum concentration at about 25km above the surface of the Earth. It is constantly being created and destroyed through natural chemical cycles. The layer is critical to life on Earth because it filters out damaging ultraviolet radiation from the Sun.

During the 1920s, chlorofluorocarbons (CFCs) were invented, and until the 1970s they were considered to be the ideal chemical for many industrial and consumer applications. Being inert, non-toxic, and cheap, they became almost indispensable for refrigeration, foam blowing, aerosol propellants, fire extinguishers, and as solvents. However, a possible problem with CFCs was recognized in the early 1970s, when it was realized that chlorine compounds (such as CFCs) produced by industry could deplete the ozone in the stratosphere.

Specifically, when a CFC molecule is released and eventually carried up into the stratosphere, it is decomposed by solar ultraviolet radiation and it produces an unattached chlorine atom. This atom can initiate an ozone-destroying reaction sequence while re-emerging unchanged itself. The single chlorine atom can emerge from the reaction sequence perhaps 100,000 times before it is finally removed by something else. Accordingly, the 100 grams of CFC in a single spray-can, or a single domestic refrigerator, can eventually destroy over 3 tonnes of ozone. The millions of tonnes of CFCs currently in the atmosphere will continue to leak into the stratosphere and affect the ozone for the next 200 years. Already, observational data from many parts of the Earth are showing a small downward trend in ozone.

ozone-depleting potential

The unit for measuring the relative impact of chlorine and bromine-based compounds on ozone. CFC-11 serves as the reference for the ozone-depleting potential, and has a value of 1.

ozone impacts

The main long-term effect of reduced ozone amounts is an increase in ultraviolet radiation near the Earth's surface. This can lead to an increase in skin cancer, the degradation of many materials such as plastics, paints, and fabrics, and the reduction in productivity of significant crops such as rice, and in oceanic plankton concentrations.

ozone layer

That layer of the atmosphere in which the concentration of ozone is greatest. The term is used both to signify the layer from about 10km to 50km in which the ozone concentration is appreciable, and also to signify the much narrower layer from about 20km to 25km in which the concentration generally reaches a maximum. The ozone layer absorbs high-energy ultraviolet-B radiation, and transforms it into heat. Reductions in the total ozone column density lead to an increase in the intensity of the harmful ultraviolet-B radiation at the Earth's surface.

ozone: tropospheric and stratospheric

The larger part of atmospheric ozone is found in the stratosphere at an altitude between 12km and 40km, where it is generated in the process of the photo-decomposition of oxygen. The ozone in the troposphere accounts for approximately one-tenth of the total ozone column. This ozone is mainly produced from hydrocarbons and nitrogen oxides in a photochemical process triggered by "smog mechanisms".

Tropospheric ozone has serious negative effects (toxic for humans, animals and plants; aggravation of the greenhouse effect), whereas stratospheric ozone is of vital importance as a filter of ultraviolet-B radiation.

OZONET

An ozone network. An international online database on CFC alternatives, run through the General Electric Information Services network.

P

pack ice

Pack ice is a term which is usually used to describe ice which covers more than half of the visible sea surface. Of the sea surface, broken, loose, or open pack ice usually covers 50–80% of the sea surface, while close, dense, thick or tight pack ice usually covers more than 80%.

Pacific Science Association (PSA)

The Association was created in 1920 at the First Pan-Pacific Scientific Conference. Its objectives are: to review and establish priorities of common scientific concerns in the Pacific Basin and to provide a multidisciplinary forum for discussion of these concerns through congresses and inter-congresses and other scientific meetings; to initiate and promote co-operation in the study of scientific problems relating to the Pacific region, more particularly those affecting the prosperity and wellbeing of Pacific peoples; and to strengthen the bonds of peace among Pacific peoples by promoting a feeling of "brotherhood" among the scientists of all the Pacific countries.

PAGES see **Past Global Changes**

palaeoclimatology

The study of climate conditions in the geological past.

palaeoecology

The study of the relationships between ancient organisms and their environments.

palynology

The study of fossil spores, and especially pollen. As spores and pollen are usually adapted to resist destruction and to be dispersed over large distances, they are valuable for use in the correlation of the rocks in which they occur. They are also important environmental indicators and have been used in monitoring climatic change during the Quaternary.

Panel on World Data Centres (Geophysical, Solar and Environmental)

The Panel on World Data Centres (Geophysical, Solar and Environmental) was established in 1968 at the 12th General Assembly of ICSU, to advise the officers of ICSU on the management of the World Data Centres. The Panel oversees the operation of about 40 World Data Centres, which are maintained by their host countries and are responsible for collecting, distributing and archiving a wide range of data. These data provide baseline information for research in many ICSU disciplines, especially for monitoring changes in the geosphere and biosphere – gradual or sudden, foreseen or unexpected, natural or manmade.

particulate matter
Very small pieces of solid or liquid matter, such as particles of soot, dust, aerosols, fumes or mists.

particulates
Fine liquid or solid particles such as dust, smoke, mist, fumes, or smog, found in the air or emissions.

Past Global Changes (PAGES)
Past Global Changes (PAGES) is an IGBP core project, which co-ordinates and integrates existing national and international palaeo-projects, and implements new activities in order to obtain information on the pre-industrial variations of the Earth system, and the baseline on which human impacts are superimposed. Typical research tasks are the separation of anthropogenic and corresponding responses, and the documentation of possible internally forced processes. Of particular interest are the deconvolution of long-term climatic changes over a glacial cycle of ecosystems to the warming at the end of the last glaciation, and changes in the atmospheric content of carbon dioxide and methane throughout the last glacial cycle, and during periods of abrupt climatic change.

Pathfinder Datasets
A NASA/NOAA project to be included in the WMO Climate Change Detection Project (CCDP).

perception studies
Perception studies recognize that the "objective" environment, as described by scientific observation, does not necessarily correspond with the environment as perceived by people. This is certainly true of climate; different individuals and cultures may see climate and its fluctuations differently. Thus, decision-makers may differ markedly in the way they respond to or plan for similar climate stresses. Perception studies in climate impact assessments attempt to identify the different mental images of climate, in order to improve understanding of human response to climate fluctuation.

Perception studies can be particularly valuable when little objective analysis of climate impacts is available. Places with little or no data on climate, crop yields, water resources or other related topics can be studied by interviewing people, observing their climate-sensitive activities, or by analyzing historical and contemporary written sources for information regarding past climate and its impacts. Carefully done, perception studies can give an assessor some idea of how people are likely to respond to future climate changes.

The goals of perception studies as part of climate impact assessment include: describing how people view climate and related aspects of the natural environment, assessing "climate expectations" and possible responses to climate fluctuation, and understanding why decision-makers respond as they do to climate effects or information on possible climate changes.

periglacial
Pertains to conditions bordering a frozen or ice-covered region (such as a glacier), especially in regard to climatic or geological processes.

perihelion
The point on the Earth's orbit which is nearest the Sun. At present, the Earth reaches this point in about mid-January. (See also **aphelion**)

permafrost
Perennially frozen ground that occurs wherever the temperature remains below 0°C for several years. The active layer refers to that portion of the ground that freezes in winter and thaws in summer, usually less than 1m in depth.

Permanent Representatives (PRs) of Members with WMO
Each Member of WMO designates a Permanent Representative (PR) who is usually the director of the meteorological or hydrometeorological service to act on technical matters for the Member between sessions of Congress. Permanent Representatives are the normal channel of communications between the WMO and their respective countries, and they maintain contact with the competent authorities, governmental or non-governmental, of their own countries on matters concerning the work of the WMO.

Permanent Service for Mean Sea Level see Federation of Astronomical and Geophysical Data Analysis Services

perturbation
Any departure introduced into an assumed steady state of a system.

pH scale

A pH scale is used to measure the degree of acidity or alkalinity of a liquid. A pH of 7 indicates neutrality. The greater the acidity, the lower the pH value; the greater the alkalinity, the higher the pH value. The scale is logarithmic, so a drop in pH from say 5 to 4 corresponds to a 10-fold increase in the acidity of the liquid. Clean rain is slightly acidic with a pH of 5.6. The rain that falls over eastern North America now has a pH of 4.6 or lower – ten times as acidic as "normal" rain.

phenology

Phenology is the study of climate conditions on the basis of the flowering, ripening and harvesting of plants and crops. It has been found that there is a close correlation between the accumulated temperature of the vegetative period and the dates of blossoming and fruiting. The first paper to bring together large amounts of phenological data – specifically on grape harvests – was published in 1883 based on French, German and Swiss wine-harvest records from the 14th century.

photochemical bioclimatology

Essentially this is the investigation of the effects of light from the Sun and sky. Sunburn of the skin and cornea of the eye is initiated by denaturization of nucleic acid and skin proteins, causing a local histamine-like action. In nature, the effect is restricted and is induced by ultraviolet radiation. Sunburn is delayed or prevented by three screening agents in the skin: pigment, horny layer and urocanic acid. The pigment of permanently brown or black races, as well as the radiation-induced pigment in variably coloured races, acts as a protective filter. The horny layer of the skin reportedly grows in thickness by ultraviolet exposure. The third screen is urocanic acid, a substance derived by enzymatic action from the amino acid L-histidine in the sweat. Frequent exposure over years leads to skin elastosis and finally skin cancer. Skin carcinomas occur more frequently on facial skin exposed to sunlight.

Most sunlight and skylight received by the eye is that reflected by clouds and surfaces. The intensity of the incoming light, its angle of incidence and the amount and kind (specular or diffuse) of albedo control the amount of light received by the eye. Many specular reflections from water, ice, metals, snow, clouds and white sand are bright enough to irritate the eye. Most sunglasses dampen the whole visible spectrum uniformly and eliminate ultraviolet and infrared rays.

photochemical oxidants

Atmospheric gases produced by photomechanical reactions in the presence of ultraviolet radiation, which are mainly composed of carbon monoxide, hydrocarbons and nitrogen oxides. One of the most important photo-oxidants is tropospheric ozone.

photochemical smog
Air pollution initiated or aided by solar radiation and caused by chemical reactions among various substances and pollutants in the atmosphere.

photosynthesis
The process by which green plants use light to synthesize organic compounds (primarily carbohydrates) from carbon dioxide and water, using light absorbed by chlorophyll as an energy source. Oxygen and water vapour are released in the process. Photosynthesis is dependent on favourable temperature and moisture conditions, as well as on the atmospheric carbon dioxide concentration. Increased levels of carbon dioxide can increase net photosynthesis in many plants.

PIE
The Polar Ice Extent Project.

pilot-balloon observations
The determination of upper winds by the optical tracking of a free balloon.

planetary boundary layer
The atmospheric boundary layer, or the layer of the atmosphere from the surface to the level where the frictional influence is absent.

plankton
Aquatic and usually microscopic organisms that float and drift passively in the water. Plankton consists mainly of animal larvae, protozoa, and plants such as diatoms. Phytoplankton refers to plant forms, on which all other marine organisms depend, directly or indirectly, for their survival; zooplankton refers to the immense variety of animal forms. The distribution of both phytoplankton and zooplankton throughout the oceans is very variable: for example, plankton-rich areas form highly fertile zones in the oceans where upwelling currents occur.

Pleistocene
The earlier of the two Quaternary epochs, extending from the end of the Pliocene, about 1.8 million years ago, until the beginning of the Holocene about 10,000 years ago. During this period, the world experienced great fluctuations in temperature, resulting in cold periods (glacials) separated by warmer periods (interglacials). In the Alps, four main glacial episodes have been recognized: the Gunz, Mindel, Riss, and Wurm, but recent research indicates that at least two other glacial episodes – the Donau and the Biber – should be recognized.

Pliocene climatic optimum
Reconstructions of summer and winter mean temperatures and total

annual precipitation in the Pliocene climatic optimum period (about 3.3 to 4.3 million years ago) have been made for this period. Many types of proxy data are used to develop temperature and precipitation patterns over the landmasses of the Northern Hemisphere. The research suggests that Northern Hemisphere summer temperatures were about 3-4°C higher than present-day values, and that atmospheric concentrations of carbon dioxide are also estimated to be twice as large as pre-industrial values of the modern era. However, some studies show lower values, and other studies claim that even in the very warm middle Eocene (about 40,000,000yr BP) atmospheric carbon dioxide concentration may have been no larger than that found today.

Such considerations cast at least some doubt on the extent to which atmospheric carbon dioxide concentrations were higher than present values during the Pliocene. Uncertainties associated with reconstructions of the Pliocene climatic optimum are considerable and include: imprecise dating of the records, especially those from the continents (uncertainties of 100,000 years or more); differences from the present day surface geography, including changes in topography; and that the ecology of life on Earth from which many of the proxy data are derived was significantly different.

pluvial period

A period of time experiencing greater rainfall than preceding or succeeding periods, usually on a geological timescale. Many of the semi-desert areas of the tropics experienced pluvials during full glaciation in polar regions. This was a response to the southward movement of the circulation belts and depression tracks. Evidence for these wetter periods comes from increased plant growth, and animal remains found in areas in which they would not be found under present conditions. However, lower temperatures could have reduced evaporation, making the same rainfall amounts more effective.

pluviograph

A raingauge which includes an arrangement for the time recording of the depth of water from precipitation.

point data

Observations at a specific geographical site, such as the site of a rain gauge or a stream-gauging station.

polar easterlies

A diffuse belt of low-level easterly winds located on the poleward side of the axis of the subpolar low-pressure belt.

pollen analysis

A technique of both relative dating and environmental reconstruction, consisting of the identification and counting of pollen types preserved in peats and lake beds.

Specifically it is the study of fossil pollen and spore assemblages in sediments, especially when reconstructing the vegetational history of an area. The outer coat of a pollen grain or spore is very characteristic for a given family, genus, or sometimes even species. It is also very resistant to decay. Thus, virtually all spores and pollen falling on a rapidly accumulating sediment, anaerobic water or peat are preserved. With careful interpretation, pollen analysis enables examination of climatic change and human influence on vegetation, as well as sediment dating and direct study of vegetation character. Much of what was known of postglacial climate history until about 1950 was contributed by pollen analysis.

potential evapotranspiration

The maximum quantity of water capable of being evaporated by a surface and transpired by vegetation, when the surface and the vegetation are well supplied with water.

positive feedback

An interaction within a system that causes an amplification of the response of the system.

precipitable water

The total atmospheric water vapour contained in a vertical column of unit cross-sectional area extending between any two specified levels, commonly expressed in terms of the height to which that water substance would stand if completely condensed and collected in a vessel of the same unit cross section. The total precipitable water is the water contained in a column of unit cross section extending all the way from the Earth's surface to the "top" of the atmosphere.

precession

The gyroscope movement of the Earth's pole around the ecliptic, due to the attraction of the Sun and Moon. One complete revolution takes approximately 20,000 years.

precipitation

Any and all forms of water, whether liquid or solid, that fall from the atmosphere and reach the Earth's surface.

precursor signals

Signals detectable above the "noise" of natural climatic variability, which indicate a significant change in a climate parameter attributed to an increase in atmospheric carbon dioxide concentrations. The signal may be estimated by numerical modelling of the climate, and the noise can be estimated using instrumental data.

prevailing wind direction
The wind direction most frequently observed during a specific time period.

primary energy
Primary energy is the sum of the raw materials used for the production of energy; that is, all natural energy sources including fossil fuels such as hard coal, lignite, petroleum, natural gas, oil shale, tar sand; nuclear fuels such as uranium and thorium; and renewable energy sources such as hydroelectric power, wind energy, solar energy, geothermal heat and biomass.

primary forest
Virgin forest. In the strict sense of the term, a self-sustaining climax forest stand that has been influenced so little by humans that its physiognomy has been essentially shaped and determined by its natural environment.

probability forecast
A forecast of the probability of occurrence of one or more of a mutually exclusive set of weather contingencies, as distinguished from a series of categorical statements.

protectionism
External trade practices aimed at protecting an entire national economy, individual industries, or certain domestic producers against foreign competition. These can take the form of tariffs, quotas, or – to an increasing extent in recent years – non-tariff trade barriers.

proxy climate indicators
Dateable evidence of a biological or geological phenomenon whose condition, at least in part, is attributable to climatic conditions at the time of its formation. Proxy data are any material that provides an indirect measure of climate; they include documentary evidence of crop yields, harvest dates, glacier movements, tree rings, varves, glaciers and snowlines, insect remains, pollen remains, marine microfauna, and isotope measurements including ^{18}O in ice sheets, ^{18}O, ^{2}H, and ^{13}C in tree rings, and $CaCO_3$ in sediments and speleothems. There are three main problems in using proxy data: dating, lag and response time, and meteorological interpretation. Tree rings, ice cores and pollen deposits from varved lakes are the most promising proxy data sources for reconstructing the climate of the past 5,000 years, since the datings are precise on an annual basis, while other proxy data sources may yield data on only a 50 ± 100 years timescale.

proxy data
Includes data such as glacier mass balance, sedimentary cores, tree rings and permafrost records.

PRs see **Permanent Representatives of Members with WMO**

PSA see **Pacific Science Association**

PSO
The Polar Stratospheric Ozone Project.

psychrometer
An instrument used to measure the humidity of the atmosphere. It comprises two identical thermometers, the bulb of one of which is dry, while that of the other is wet and covered by a film of pure water or ice.

pyranometer
Instrument for measuring the solar radiation received on a plane surface.

Q

Quarterly Bulletin on Solar Activity see **Federation of Astronomical and Geophysical Data Analysis Services**

quasi-biennial oscillation
A stratospheric oscillation of equatorial east–west winds which vary with a period of 26–30 months. A typical oscillation shows an easterly flow for 12–16 months, and a reversal to a westerly flow for 12–16 months.

Quaternary
The period of geological time following the Tertiary Period. It is formed of two epochs, the Pleistocene and Holocene, and it extends from 1.8 million years ago until the present. During this time, parts of Europe were subjected to four major advances of the ice sheets, which were separated by warmer interglacial episodes. The land fauna alternated between forms adapted to cold conditions, such as the mammoth and woolly rhinoceros, and species now restricted to the tropics. Pollen and the remains of beetles have proved valuable in monitoring climatic change. Human beings became the dominant terrestrial species.

quotas
The quantitative limits placed on the importation or exportation of specified commodities.

R

radiation

All bodies emit radiation, which is electromagnetic energy. The total energy of the radiation emitted depends on the fourth power of the temperature, so that hot "bodies" emit much more radiation than colder bodies. The principal source of radiation is the Sun, and radiation is often considered to be the portion of the electromagnetic radiation spectrum emitted by the Earth and Sun. In terms of wavelength, this is radiation encompassing part of the ultraviolet, all the visible, and part of the infrared spectrum. The Sun has an emission temperature of 6000°K giving an energy maximum in the visible-light wavelength. This reaches the "top" of the atmosphere at the rate of 1.35kW per square metre. On its passage through the atmosphere, some of this shortwave radiation is reflected back to space by clouds and dust, some is scattered by gas molecules and dust particles to give diffuse radiation, and some is absorbed by water vapour, carbon dioxide and dust. The remainder reaches the ground surface where some is reflected and the rest is absorbed.

The Earth's surface has a mean emission temperature of about 288°K as a result of solar heating. Thus terrestrial radiation is in the longer wavelengths. The gases of the atmosphere (especially water vapour, and several greenhouse gases) have a very different response to this radiation and much of it is absorbed. The atmosphere warms as a result of this absorption, and atmospheric counter-radiation is returned to the surface and helps to maintain higher temperatures than would otherwise be expected.

radiation balance

The difference between incoming and outgoing radiation at any point. Normally there is a surplus of radiation on the ground surface during the day, which helps to warm the surface and atmosphere, and a deficit at night when cooling takes place. Taking the Earth and atmosphere together, the areas equatorwards of 38° latitude have a radiation surplus, while polewards of 38° latitude there is a radiation deficit.

radiation flux

The amount of radiation impinging on a given surface per unit time.

radiative forcing

There are several natural factors which can change the balance between the energy absorbed by the Earth and that emitted by it in the form of longwave infrared radiation; these factors cause the radiative forcing on climate. The most obvious of these is a change in the output of energy from the Sun. There is direct evidence of such variability over the 11-

year solar cycle, and longer-period changes may also occur. Slow variations in the Earth's orbit also affect the seasonal and latitudinal distribution of solar radiation. One of the other most important factors is the greenhouse effect.

The major contributor to increases in radiative forcing due to increased concentrations of greenhouse gases since pre-industrial times is carbon dioxide (61%), with substantial contributions from methane (17%), nitrous oxide (4%) and chlorofluorocarbons (12%). Stratospheric water vapour increases, which are expected to result from methane emissions, contribute 6%, although evidence for changes in concentration is based entirely on model calculations.

The most recent decadal increase in radiative forcing is attributable to carbon dioxide (56%), methane (11%), nitrous oxide (6%), CFCs (24%) and stratospheric water vapour (4%). Stratospheric aerosols resulting from volcanic eruptions can cause a significant radiative forcing. A large eruption, such as El Chichon, can cause a radiative forcing, averaged over a decade, about one-third of (but the opposite sign to) the greenhouse gas forcing between 1980 and 1990. Regional and short-term effects of volcanic eruptions can be even larger.

radiatively active gases

Gases that absorb incoming solar radiation or outgoing infrared radiation, thus affecting the vertical temperature profile of the atmosphere. Gases most frequently being cited as being radiatively active are water vapour, carbon dioxide, methane, nitrous oxide, chlorofluorocarbons and ozone.

radiocarbon dating

A dating method for organic material that is applicable to about the past 70,000 years. It relies on the known rate of decay of radioactive carbon, of which half is lost in a period (the "half-life") every 5730 ± 30 years. In principle, since plants and animals constantly exchange carbon dioxide with the atmosphere, the ^{14}C content of their bodies when alive is a function of the radiocarbon content of the atmosphere. When an organism dies, this exchange ceases and the radiocarbon fixed in the organism decays at the known half-life rate. Comparison of residual ^{14}C activity in fossil organic material with modern standards enables the age of the samples to be calculated.

radiometer

An instrument for measuring radiation.

radiosonde

An instrument carried into and through the atmosphere, equipped with devices permitting several meteorological elements (pressure, temperature, humidity, etc.) to be measured, and which is provided with a radio transmitter for sending this information to a receiver.

radiosonde observation
The observation of meteorological elements in the upper air, usually atmospheric pressure, temperature, and humidity, by means of a radiosonde from which the readings are transmitted instantaneously to a receiving station.

radiowind observation
The determination of upper winds by the tracking of a free balloon by electronic means.

rain
Precipitation of liquid water particles, either in the form of drops of more than 0.5mm diameter or of smaller widely scattered drops.

raingauge
An instrument for measuring the amount of rain; specifically, the depth of water from precipitation over a horizontal impervious surface that is not subject to evaporation.

rain shadow
A region, situated on the lee side of a mountain or mountain range, where the rainfall is much lower than on the windward side.

rainforest
A term used loosely for forests of broad-leaved, mainly evergreen, trees found in continually moist climates in the tropics, subtropics, and some parts of the temperate zones. The tropical rainforest includes the vast Amazon forest as well as large areas in western and central Africa, Malaysia, Indonesia,and New Guinea. Estimates of the total world area of rainforest vary from $5,500,000km^2$ to $9,400,000km^2$. Tropical rainforests, unlike temperate forests of beech or oak, are usually mixed in composition, with no one species forming a large proportion of the whole stand. In some parts of the tropics, however, there are rainforests dominated by a single species.

rawinsonde
A radiosonde which is tracked by radar, thus enabling the winds in the upper atmosphere to be determined.

rawinsonde observation
A combined radiosonde and radiowind observation.

real-time processing
The concurrent processing of data upon their reception.

real-time transmission
The transmission of data immediately upon their observation.

Re-analysis Project
A project of the US National Center for Atmospheric Research (NCAR) which involves a re-analysis of all available past data to achieve an homogeneous historical gridded dataset.

reef
A reef may be formed of solid rock or pebbles, but the term is more commonly applied to organic reefs. Coral, a hard calcareous material, dead or alive, is usually predominant in all organic reefs, and coral is the commonest reef-building organism.

reference climate network see reference climatological station

reference climatological station
A climatological station, the data of which are intended for the purpose of determining climatic trends. This requires long periods (ideally 30 years or more) of homogeneous records, where human-induced environmental changes have been and/or are expected to remain at a minimum.

reforestation
Activities or processes which result in the regrowth of forests removed by such practices as harvesting or slash burning.

Regional Climate Changes by 2030 (IPCC WG I: Policy-makers Summary)
The following assumes the IPCC Business-as-Usual scenario and changes from pre-industrial.

The numbers given are based on high-resolution models, scaled to be consistent with our best estimate of global mean warming of 1.8°C by 2030. For values consistent with other estimates of global temperature rise, the numbers should be reduced by 30% for the low estimate, or increased by 50% for the high estimate. Precipitation estimates are also scaled in a similar way. Confidence in the following regional estimates is low.

Central North America The warming varies from 2°C to 4°C in winter and 2°C to 3°C in summer. Precipitation increases range from 0% to 15% in winter, whereas there are decreases of 5% to 10% in summer. Soil moisture decreases in summer by 15% to 20%.

South Asia The warming varies from 1°C to 2°C throughout the year. Precipitation changes little in winter and generally increases throughout the region by 5–15% in summer. Summer soil moisture increases by 5–10%.

Sahel The warming ranges from 1°C to 3°C. Area mean precipitation increases and area mean soil moisture decreases marginally in summer. However, there are areas of both increase and decrease in both parameters throughout the region.

Southern Europe The warming is about 2°C in winter and varies from 2°C to 3°C in summer. There is some indication of increased precipitation in winter, but summer precipitation decreases by 5–15%, and summer soil moisture decreasing by 15–25%.

Australia The warming ranges from 1°C to 2°C in summer and is about 2°C in winter. Summer precipitation increases by around 10%, but the models do not produce consistent estimates of the changes in soil moisture. The area averages hide large variations at the subcontinental level.

relative humidity
The ratio of the actual amount of water vapour in the air to the amount which would be present if the air were saturated at the same temperature.

remote sensing
The measurement or acquisition of information on some property of an object or phenomenon by a recording device that is not in physical or intimate contact with the object or phenomenon under study.

renewable energy
Natural sources of energy whose supply can be replaced as they are used (e.g. hydro-power, biomass, wind, solar).

re-radiation
The process by which a body absorbs electromagnetic radiation, converts it to thermal kinetic energy and thereby assumes a higher temperature, and then radiates some of the absorbed energy as longer-wave electromagnetic radiation.

Research and Development Programme of WMO
The Research and Development Programme of WMO promotes atmospheric research, placing highest priority on weather-prediction research on all timescales. It includes the Tropical Meteorology Research Programme relating to the study of monsoons, tropical cyclones, rain-producing tropical weather systems, and droughts, a programme related to the research on and monitoring of various of environmental pollution, and a Cloud Physics and Weather Modification Research Programme.

residence time
The mean length of time during which a pollutant remains in the atmosphere.

response function

The mathematical representation of the behaviour of a system due to a specified input.

response of the "slow climate system"

The components of climate systems can be considered to be either "fast" or "slow". "Fast" systems include the atmosphere, the upper ocean, and the land surface which have "adjustment" times of less than a few years, whereas the "slow" systems include the ocean (except the upper ocean) and the ice sheets which "adjust" on timescales of decades to centuries.

Climate model studies suggest that, if the fast system alone were involved in climate change, the Earth's climate would almost immediately follow the computed increase in greenhouse heating. Computations made with these simplified climate models indicate that the rise in greenhouse gases concentration already caused by human activities since the eighteenth century corresponds to a global surface warming in excess of 1.5°C. Such a climate change has not occurred yet; indeed research indicates that temperatures during the period 1860–80 were essentially the same as today, although a noticeable cooling and subsequent warming occurred in the intervening period. However, combining measurements over ocean and land indicates a warming trend of about 0.6°C over the past hundred years. Most of this temperature rise was observed during the first fifty years from 1880 to 1930, in spite of the fact that less than one third of the total greenhouse gas increase occurred during that period. Global warming for the period 1940–90 has been about 0.2°C. It is clear, therefore, that the real climate is behaving quite differently from the fast response of a planet with only a very shallow ocean, and does not follow directly the trend in radiative heating induced by the increasing greenhouse effect.

The response of the slow climate system is therefore paramount in any climate change discussions. The "slow system" is controlled by two basic parameters: first, the change in the net energy flux at the surface of the ocean produced by the increase in the concentration of greenhouse gases and subsequent readjustments of the atmosphere and land surface; secondly, the heat capacity of that part of the ocean which will be effectively subject to warming over a period of several decades. The effective heat capacity of the ocean is determined by the penetration of heat in deep water. This penetration into the upper kilometre of the ocean cannot be represented by a simple diffusion process because it results, in the main, from the sinking of water at high latitudes and subsequent equatorward and downward circulation in the ocean.

retrogression

The motion of an atmospheric wave or pressure system in the direction opposite to that of the basic flow in which it is embedded.

return period
The probable time period between the repetition of extreme events.

roughness
A measure of the resistance encountered by air flowing over the Earth's surface.

S

Saarbrucken International Conference (1990)
Highlights of deliberations of the Saarbrucken International Conference on Energy in Climate and Development include the following:

The conference stressed the need to quantify the target to reduce greenhouse gas emissions, particularly through an international convention. Such a convention should aim at a 20% reduction of global CO_2 emissions by the year 2005 as set by the Toronto Conference 1988. A target at this level requires a minimum reduction of 30% of CO_2 emissions in industrialized countries while facilitating energy-efficient development in the developing countries. Other greenhouse gas emissions such as methane, nitrous-oxides, etc., should be included in such an international convention.

There were several points of common agreement among the participants. The chief consensus was that issues of energy and climate change are global in nature and any attempt to resolve them would require the co-operation of all the countries. At the same time it was recognized that effective co-operation will be possible only if the proposed remedial measures take into account the vastly differing socio-economic situations among groups of countries. Each country would have to interpret the energy options and seek national solutions on its own.

There are as yet no agreed ways by which the global environmental concerns could be resolved while at the same time protecting the legitimate interests and aspirations of the developing countries. There seems no alternative to facilitating widespread and easy access and adaptation of environmentally sound technologies. Market mechanisms alone may not be able to achieve all of this. Neither will the expectations for the free transfer of technologies be realizable under present conditions. A middle ground has to be sought in the coming years through the available negotiation mechanisms.

There was no difference of opinion on the need to internalize the environmental costs in energy prices. Suggestions were made on the need to develop criteria for assessing the environmental costs. Possibilities of shifting tax burdens to increase the price of environmentally unfriendly energy systems were suggested. Increased investment in accelerated development and use of renewable energy technologies was proposed.

SAC see **Scientific Advisory Committee for the WCIRP**

SAGE
The Stratospheric Aerosol and Gas Experiment.

salinity
A measure of the concentration of dissolved salts, mainly sodium chloride, in saline water and sea water.

saltwater intrusion
A phenomenon occurring when a body of salt water invades a body of fresh water. It can occur in either surface- or groundwater bodies.

satellite (geostationary)
A satellite whose high-altitude orbit of about 36,000km is in the equatorial plane, and which thus has an orbital velocity matching that of the Earth, so that its position remains constant with respect to the Earth.

satellite (polar orbiting)
A satellite circulating the Earth once every 1–2 hours whose orbit is approximately Sun-synchronous, with a low altitude of about 1,000km. The satellite passes close to the Poles on each orbit.

savanna
A vegetation type of the semi-arid tropics, consisting of grasslands dotted with isolated trees or woody patches. With increasing moisture, the area covered by wood vegetation grows.

SCAR see **Scientific Committee on Antarctic Research**

scattering
The process by which small particles suspended in a medium of a different index of refraction diffuse a portion of the incident radiation in all directions.

scenario
Description of a possible condition, based on certain assumptions. The results of scenarios (unlike forecasts) depend on the boundary conditions of the scenario.

Scientific Advisory Committee for the WCIRP (SAC)
The Scientific Advisory Committee of the World Climate Impact

Assessment and Response Strategies Programme (WCIRP) was established by UNEP in 1980 to advise on matters relating to the programme and budget of the WCIRP. It normally meets once a year.

Scientific Committee for Biotechnology (COBIOTECH)

ICSU's Scientific Committee on Biotechnology (COBIOTECH) was created in 1986 to promote biotechnology for the benefit of humankind and to provide information and advice on biotechnology for the international community as a whole.

Scientific Committee on Antarctic Research (SCAR)

This Committee was formed at The Hague in March 1958. The main purpose of SCAR is to provide a forum for scientists of all countries with research activities in the Antarctic to discuss their field activities and plans and to promote collaboration between them.

Scientific Committee on Oceanic Research (SCOR)

This Committee was established by the Executive Board of ICSU in July 1957 to further international scientific activity in all branches of oceanic research. To do so, SCOR examines problems and identifies elements that would benefit from enhanced international action, including improvement of scientific methods, design of critical experiments and measurement programmes, and relevant aspects of science policy. SCOR also fosters recognition of individual marine scientists and laboratories, presents the views of marine scientists to the appropriate international community, and co-operates with national and international organizations concerned with scientific aspects of ocean research and interrelated activities.

Scientific Committee on Problems in the Environment (SCOPE) of ICSU

SCOPE has pursued work on biogeographical cycles in relation to the global climate cycles (SCOPE publications 13, 16 and 23), and more specifically in relation to climate-change impacts (SCOPE publication 29).

Scientific Committee on Solar–Terrestrial Physics (SCOSTEP)

This committee was originally established as an Inter-Union Commission following a Resolution adopted at the 11th General Assembly of ICSU in Bombay in January 1966. It was modified by the 14th General Assembly of ICSU in Helsinki in 1972 to a Special Committee, and by the 17th General Assembly in 1978 to a Scientific Committee. Its principal tasks are: to promote international interdisciplinary programmes in solar–terrestrial physics; to define the data relating to these programmes that should be exchanged through the World Data Centres; to provide such advice as may be required by the ICSU bodies and World Data Centres concerned with these programmes; and to work with other ICSU bodies in the co-ordination of symposia in solar–terrestrial physics, especially on topics related to SCOSTEP's programmes.

SCOPE see **Scientific Committee on Problems in the Environment**

SCOR see **Scientific Committee on Oceanic Research**

SCOSTEP see **Scientific Committee on Solar–Terrestrial Physics**

scrubber
A device that uses a liquid spray to remove aerosol and gaseous pollutants from an airstream. The gases are removed by either absorption or chemical reaction. Solid and liquid particulates are removed through contact with the spray. Scrubbers are used for both the measurement and control of pollution.

sea-level changes
In past historical times, variations of the mean sea level at a particular shore have been caused mainly by tectonic effects which force the land to sink or rise locally with respect to the global reference geoid. If all the glaciers and small inland ice caps were to disappear entirely and add their meltwater to the ocean, they would cause the global sea level to rise by 33cm. On the other hand, the interplay between snow accumulation on large ice caps and melting or iceberg calving, when ice discharge takes place into the sea, could in principle cause much more substantial changes in global sea level.

Ice budget estimates for the Greenland and Antarctic ice caps under the warmer climate conditions expected in the 21st century show that increased ablation is expected to dominate the ice budget of the Greenland ice cap, yielding a general decrease of the inland ice mass and a 13–26cm rise in mean sea level due to the release of meltwater during the next century. In contrast, temperatures are and will remain well below freezing in Antarctica in the foreseeable future so that melting will not take place.

The global ocean, on the other hand, has been storing energy for several decades and may continue to do so at an even faster rate during the next century as the greenhouse effect increases. Variations of the thermal expansion coefficient are relatively moderate in the upper layer of the ocean, where most of the warming will take place, accordingly the linear expansion of a water column is, to a first order, independent of its depth and directly proportional to the total amount of heat stored in the column.

The thermal expansion of the oceans is therefore related to estimating the integrated heat intake of the world ocean over a particular period in the future. Based on a full three-dimensional computation of the initial response of a coupled ocean–atmosphere model to doubling the concentration of carbon dioxide, a net heating rate of 2.5 watts/km^2 has been suggested. This value has been obtained after adjustment of model properties in the troposphere and on the land surface (the fast component of the climate system), but before significant adjustment of the ocean temperature (the slow component). But, since the ocean surface must warm up eventually, it is plausible that the imbalance, measured by the net energy flux at the ocean/atmosphere interface, should not exceed 2.5 watts/km^2 during the next century. This line of reasoning places an upper limit of about 40cm on the global sea-level rise which can be expected from thermal expansion of the oceans by the year 2100.

sea-level changes (human-induced)

There is a general consensus that human-induced global warming will be accompanied by a rise in global mean sea level. The "best" estimate is that this will lie between about 17cm and 26cm by 2030, corresponding to the 1–2°C warming over the same period. Given the full range of uncertainties, the rise could be as little as about 5cm or as large as about 45cm (but these extremes are unlikely). The main factors changing the ocean water volume are likely to be the melting of mountain glaciers and the expansion of the warming seas. Changes in Antarctic ice sheets, where about 90% of the Earth's land ice is located, are unlikely to contribute to sea-level rise on this timescale and in fact may have a small negative influence due to increased precipitation and ice accumulation. Greenland ice sheets are likely to make a small positive contribution to the rise. However, because of the slow process of heat transfer from atmosphere to the ocean, and the very long response times of polar ice, even if global warming stopped abruptly in 2030, global sea level would continue to rise for many decades, and possibly for hundreds or even many hundreds of years.

Indeed, on many of the world's coasts sea level is still changing for reasons other than climate change. In Scandinavia, for example, the land is slowly rising, following the disappearance of ice sheets some 10,000 years ago, and relative sea level is declining in some areas at a rate of 10cm/decade. Some coastal areas are also sinking; for example, in southeast England, the relative sea level is rising by about 2.5cm per decade – twice the global average. In addition, some tropical islands have been built up by coral growth on the foundations of extinct volcanoes which are slowly subsiding.

sea-level measurements

Few coastal stations exist where long time-series of consistent sea-level measurements are available, and these stations are not distributed evenly around the globe. For this reason, estimates of variation in the global mean sea level can be obtained only for the past century. The most rigorous analysis of tide-gauge data indicates a global rising trend at the rate of 1.4cm per decade, with a tendency to accelerate in the past fifty years.

sea-level rise

The IPCC Scientific Assessment Report states that despite many problems associated with estimates of globally coherent secular changes in sea-level measurements based on tide gauge records, we conclude that it is highly likely that sea level has been rising over the past 100 years.

The Report further states the for the IPCC Business-as-Usual Scenario at year 2030, global mean sea level is expected to be 8–29cm higher than today, with a best estimate of 18cm. At the year 2070, the rise is expected to be 21–71cm, with a best estimate of 44cm.

sea-surface temperature

The temperature of the "upper" layer of sea water (approximately 0.5m deep). The accurate measurements of sea-surface temperatures is difficult and, because of the different measuring techniques, it is at times quite difficult to compare measurements of sea-surface temperatures both spatially and over time.

Second International Polar Year (1932–33)

The second International Polar Year in 1932–33 placed a new emphasis on studying the extent to which observations in the polar regions could improve the accuracy of weather forecasts in other parts of the world. Additionally, it tackled the problem of how a better knowledge of meteorological conditions at high latitudes would help sea and air transport (the possibility of air routes across the Arctic had already been raised at that time). Altogether 44 nations participated in the Polar Year and a vast amount of observational data was collected. New meteorological stations were also established in the equatorial zone, but permanent stations were yet to be established in the Antarctic. The experience gained in processing and archiving these data led to a later proposal by ICSU for the creation of World Data Centres.

Second World Climate Conference (SWCC)

The Second World Climate Conference was convened in Geneva, Switzerland, from 29 October to 7 November 1990, under the sponsorship of WMO, UNEP, UNESCO and its IOC, FAO and ICSU. A Conference Statement was adopted by the participants in the scientific and technical sessions from 29 October to 3 November 1990, on the basis of the presentations at the Conference, the deliberations of task groups of participants organized to address various specific issues, and plenary discussions involving all participants. The scientific and technical sessions involved 747 participants from 116 countries. The following is the official Summary of the Conference Statement:
* Climate issues reach far beyond atmospheric and oceanic sciences, affecting every aspect of life on this planet. The issues are increasingly pivotal in determining future environmental and economic wellbeing. Variations of climate have profound effects on natural and managed systems, the economies of nations and the wellbeing of people

everywhere. A clear scientific consensus has emerged on estimates of the range of global warming which can be expected during the 21st century. If the increase of greenhouse gas concentrations is not limited, the predicted climate change would place stresses on natural and social systems unprecedented in the past 10,000 years.

* At the First World Climate Conference in 1979, nations were urged "to foresee and to prevent potential manmade changes in climate that might be adverse to the wellbeing of humanity". The Second World Climate conference concludes that, notwithstanding scientific and economic uncertainties, nations should now take steps towards reducing sources and increasing sinks of greenhouse gases through national and regional actions, and negotiation of a global convention on climate change and related legal instruments. The long-term goal should be to halt the build-up of greenhouse gases at a level that minimizes risks to society and natural ecosystems. The remaining uncertainties must not be the basis for deferring societal responses to these risks. Many of the actions that would reduce risk are also desirable on other grounds.

* A major international observational and research effort will be essential to strengthen the knowledge base on climate processes and human interactions, and to provide the basis for operational climate monitoring and prediction.

On 6-7 November 1990, the Ministerial Sessions of the Conference were held. A Ministerial Declaration of the Second World Climate Conference was agreed to and the following items were included in 31-paragraph Declaration:

* We note that, while climate has varied in the past and there is still a large degree of scientific uncertainty, the rate of climate change predicted by the Intergovernmental Panel on Climate Change (IPCC) to occur over the next century is unprecedented. This is due mainly to the continuing accumulation of greenhouse gases, resulting from a host of human activities since the industrial revolution, hitherto particularly in developed countries. The potential impact of such climate change could pose an environmental threat of an up-to-now unknown magnitude, and could jeopardize the social and economic development of some areas. It could even threaten survival in some small island states and in low-lying coastal, arid and semi-arid areas.

* Recognizing climate change as a common concern of mankind, we commit ourselves and intend to take active and constructive steps in a global response, without prejudice to sovereignty of states.

* Recognizing that climate change is a global problem of unique character and taking into account the remaining uncertainties in the field of science, economics and response options, we consider that a global response, while ensuring sustainable development of all countries, must be decided and implemented without further delay, based on the best available knowledge such as that resulting from the IPCC assessment. Recognizing further that the principle of equity and

the common but differentiated responsibility of countries should be the basis of any global response to climate change, developed countries must take the lead. They must all commit themselves to actions to reduce their major contribution to the global net emissions and enter into and strengthen co-operation with developing countries to enable them adequately to address climate change without hindering their national development goals and objectives.

* We reaffirm that, in order to reduce uncertainties, to increase our ability to predict climate and climate change on a global and regional basis, including early identification of as yet unknown climate-related issues, and to design sound response strategies, there is a need to strengthen national, regional and international research activities in climate, climate change and sea-level rise.

* We recognize that commitments by governments are essential to sustain and strengthen the necessary research and monitoring programmes and the exchange of relevant data and information, with due respect to national sovereignty. We stress that special efforts must be directed to the areas of uncertainty as identified by the IPCC.

* We maintain that there is a need to intensify research on the social and economic implications of climate change and response strategies.

* We urge that special attention be given to the economic and social dimensions of climate and climate-change research.

* In order to achieve sustainable development in all countries and to meet the needs of present and future generations, precautionary measures to meet the climate challenge must anticipate, prevent, attack or minimize the causes of, and mitigate the adverse consequences of, environmental degradation that might result from climate change.

* We urge developed countries, before the 1992 UN Conference on Environment and Development, to analyze the feasibility of and options for, and, as appropriate in light of these analyses, to develop programmes, strategies and/or targets for a staged approach for achieving reductions of all greenhouse gas emissions not controlled by the Montreal Protocol, including carbon dioxide, methane and nitrous oxide, over the next two decades and beyond.

secondary burner

A burner installed in the secondary combustion chamber of an incinerator to maintain a minimum temperature and to complete the combustion of incompletely burned gases.

secondary energy

Secondary energy sources are those sources produced by the conversion of primary energy sources in a conversion process (e.g. in refineries or power plants). These include coal products such as coke and briquettes, petroleum products such as petrol and fuel oil, gas products such as city gas and refinery gas, as well as electricity and district heating.

sedimentation
The process of settling and depositing by gravity of suspended matter in water.

selective logging
A type of logging in which only part of the trees are extracted, as opposed to clear-felling, in which all of the trees of a forest are removed in a single pass.

shifting cultivation
A type of agriculture in which cultivators periodically clear a new piece of forest in order to grow crops on it. This is necessary, since the soils of such areas are rapidly depleted and then no longer produce adequate yields. This type of agriculture is widely practised in tropical forests. In general, forest grows back on the cleared plots after they have been abandoned, and several years or decades later they are then cleared again.

shortwave radiation
The radiation received from the Sun and emitted in the spectral wavelengths shorter than 4μm. It is also called "solar radiation".

SEI
The Stockholm Environment Institute.

sensible temperature
Heat is produced constantly by the human body at a rate depending on muscular activity. For body-heat balance to be maintained, this heat must be dissipated by conduction to cooler air, by evaporation of perspiration into unsaturated air, and by radiative exchange with surroundings. Air motion (wind) affects the rate of conductive and evaporative cooling of skin, but not of lungs. Radiative losses occur from bare skin and clothing, and depend on its temperature and that of surroundings, as well as sunshine intensity.

Specifically, the so-called "sensible temperature" is the temperature at which air with some standard humidity, motion and radiation would provide the same sensation of human comfort as existing atmospheric conditions. Of the many sensible temperature formulas proposed, none is completely satisfactory or generally accepted. Most are intended for warm, moist conditions; a few, like the "wind chill" index, are for cold weather. Some are purely empirical, modifying the actual temperature according to the humidity; others are theoretical, and express estimated heat loss rather than an equivalent temperature.

Siberian high
An area of high pressure which forms over Siberia in the winter, and

which is particularly apparent on mean charts of sea-level pressure. It is centered near Lake Baikal, where the average sea-level pressure exceeds 1030 hectopascals from late November to early March. This high-pressure area is enhanced by the surrounding mountains which prevent the cold air from flowing away readily.

signal-to-noise ratio
A quantitative measure of the statistical detectability of a signal, expressed as a ratio of the magnitude of the signal relative to the variability. For first detection of climate change induced by carbon dioxide, the model signal is the mean change or anomaly in some climatic variable, usually surface-air temperature, attributed by a numerical model to increased concentrations of carbon dioxide. Observed noise is the standard deviation or natural variability computed from observations of that variable and adjusted for sample size, auto-correlation and time averaging.

slow climate system see response of the "slow climate system"

smog
Originally, a contraction of "smoke" and "fog", which characterized air pollution episodes in London, Glasgow, Manchester and many other cities. The London Smog of 1952 caused 4,000 excess deaths and led to the enactment of the Clean Air Acts of 1956 and 1968, which provided a ban on the use of smoky fuels within specified areas. The word smog has since been applied to other air-pollution effects not necessarily connected with smoke, such as the "Los Angeles smog" which arises from the photochemical action of sunlight on nitrogen oxides and hydrocarbons emitted by motor vehicles.

smoke
Aerosols of minute solid or liquid particles (most less than $1\mu m$ in diameter) formed by the incomplete combustion of a fuel. In air pollution it is mainly associated with the burning of coal.

snow
Precipitation consisting of white or translucent ice crystals and often agglomerated into snowflakes.

snowgauge
An apparatus designed to measure the amount of water that has fallen in the form of snow.

soil moisture
The moisture contained in the portion of the soil which is above the water table, including water vapour, which is present in the soil pores.

soil moisture content
The percentage of water in the soil, expressed on a dry-weight basis or by volume.

soil salinization
A process through which the concentration of soluble salts in a soil can increase as water is removed from the soil by evaporation and transpiration.

solar constant
The flux density of solar radiation (radiative energy per unit of time and area) passing through a surface perpendicular to the radiation of the Sun, outside the atmosphere of the Earth at a distance of 150 million km. It appears that the value is almost constant (between 1870 and 1990 it varied from about 1,367 watts/m^2 to 1,368 watts/m^2) except in the shortest part of the spectrum where the amount of energy received is small. The view among most solar physicists is that the total radiation reaching the "top" of the Earth's atmosphere from the Sun does not vary over time by more than a tiny fraction. However, changes in the internal processes of the Sun can influence its energy output, and some scientists suggest that it causes climatic changes on Earth.

In terms of direct effects on climate, the total solar irradiance (the so-called "solar constant") is significant. Continuous, space-borne measurements of total irradiance have been made since 1978. These have shown that, on timescales from days to a decade, there are irradiance variations associated with activity in the Sun's outer layer, the photosphere – specifically, sunspots and bright areas known as faculae. The very high-frequency changes are too rapid to affect the climate noticeably. However, there is a lower-frequency component that follows the 11-year sunspot cycle which may have a climatic effect. It has been found that the increased irradiance due to faculae more than offset the decreases due to the cooler sunspots; consequently, high sunspot numbers are associated with high solar output.

Over the period 1980–86, there was a decline in irradiance of about 1 watt/m^2 corresponding to a globally averaged forcing change at the top of the atmosphere of a little less than 0.2 watt/m^2. This change can be compared with the greenhouse forcing which, over the period 1980–86, increased by about 0.3 watt/m^2. However, over longer periods these solar changes would have contributed only minimally towards offsetting the greenhouse effect on global mean temperature because of the different timescales on which the two mechanisms operate. Indeed, because of oceanic thermal inertia, and the relatively short timescale of the forcing changes associated with the solar cycle, only a small fraction of possible temperature changes due to this source can be realized. In contrast, the sustained nature of the greenhouse forcing allows a much greater fraction of the possible temperature change to be realized, so that the greenhouse forcing dominates.

solar cycle
The periodic change in the number of sunspots. On average, the interval between successive minima of sunspot numbers is about 11 years.

solar flare
A solar explosion, unpredictable in nature and up to a few hours in duration, from a restricted region of the chromosphere of the Sun. Solar flares occur above certain types of sunspots.

solar radiation
The total electromagnetic radiation emitted by the Sun. To a first approximation, the Sun radiates as a black body at a temperature of about 5700°K; hence about 99.9% of its energy output falls within the wavelength interval from 0.15μm to 4.0μm, with peak intensity near 0.47μm. About one-half of the total energy in the solar beam is contained within the visible spectrum from 0.4μm to 0.7μm, and most of the other half lies in the near infrared, with a small additional portion lying in the ultraviolet.

solar spectrum
That part of the electromagnetic spectrum occupied by the wavelengths of solar radiation.

sources and sinks of greenhouse gases
The addition to, or removal from, the atmosphere of greenhouse gases. A climate convention definition of this term might best refer only to those sources and sinks caused by "human" activities and therefore not part of the undisturbed natural cycles.

Southern Oscillation
A fluctuation of the intertropical atmospheric circulation, in particular in the Indian and Pacific Oceans, in which air moves between the southeast Pacific subtropical high and the Indonesian equatorial low, driven by the temperature difference between the two areas. The general effect is that, when pressure is high over the eastern Pacific Ocean, it tends to be low in the eastern Indian Ocean, and vice versa. The phenomenon is strongly linked to El Niño.

The fluctuation in the intertropical atmospheric and hydrodynamical circulations manifests itself as a quasi-periodic (2–4 year) variation in sea-level pressure, surface wind, sea-surface temperature and rainfall over a wide area of the Pacific Ocean. Traditionally, Darwin and Tahiti have been used as key observational sites for assessing the magnitude of the southern oscillation.

SPREP
The South Pacific Regional Environmental Programme.

SST
The sea-surface temperature.

stakeholder
An individual, group, institution, or government with an interest or concern, either economic, societal, or environmental, in a particular measure, proposal or event.

START see **System for Analysis, Research and Training**

station documentation (of climatological stations) .
Geographical and administrative information such as: the official name; latitude, longitude and elevation; name and mailing address of the observer or co-operating agency making the observations; the official observing programme; the hours at which observations are normally taken; and advice on whether the station is considered homogeneous with nearby stations.

statistical–dynamical models
Climate computer programmes which calculate simplified climate models based on versions of the conservation equations averaged over longitude, with the effects of the synoptic eddies parameterized statistically in the meridional plane.

stratosphere
That layer of the atmosphere above the troposphere, and below the mesosphere, characterized by relatively uniform temperatures and horizontal winds. Its lower limit varies from about 8km to 20km; its upper limit is at around 45km. The base of the stratosphere makes an upper limit to the general turbulence and convective activity of the troposphere; thus, air motion within the stratosphere is largely horizontal.

stochastic model
A mathematical model based on probabilities, where the prediction of the model is not a single fixed number but a range of possible numbers. The opposite is a deterministic model.

stomata
Pores in the epidermis of the leaves of plants.

succession
The sequence of plant formations that develop over the course of time on a given site.

Sun

A luminous gaseous sphere around which the Earth moves in a slightly elliptical orbit. The radiation emitted from the Sun's luminous disc (photosphere) corresponds to a black-body radiation temperature of about 5,700°K, the internal gases being at a temperature of many millions of degrees. The energy output of the Sun varies with time. This so-called "solar activity" is associated with disturbances which are observed in the photosphere and solar atmosphere and which are in large measure inter-related. Chief among the solar disturbances are sunspots and solar flares. The relationships between solar activity and various geophysical phenomena to which they give rise are termed solar–terrestrial relationships.

sunspot

A relatively dark, sharply defined region on the solar disc, marked by an umbra approximately 2,000°K cooler than the effective photospheric temperature, surrounded by a less dark but also sharply bounded penumbra. The average sunspot diameter is about 3,700km, but can range up to 245,000km. Most sunspots are found in groups of two or more, but they can occur singly. The quantitative definition of sunspot activity is called the Wolf sunspot number, denoted R. The Wolf sunspot number is also referred to as "Wolfer sunspot number," "Zurich relative sunspot number", or "relative sunspot number".

Sunspots were first discovered in the early 17th century, and by the mid-19th century astronomers had discovered that the number of sunspots followed a cycle which averaged 11 years but varied from about 8 years to 15 years. Early this century it was determined that sunspots are in fact giant magnetic fields one thousand times as strong as the Earth's magnetic field covering areas on the Sun that are larger than the Earth. These magnetic fields are created by motions of the electrically charged particles that compose the gaseous atmosphere of the Sun, and produce the dark areas that we see as sunspots by blocking the flow of hot, luminous gas from the interior to the surface of the Sun. In recent years it has been established that the large-scale magnetic field of the Sun reverses itself each time the number of sunspots reaches a maximum. There is therefore a 22-year double sunspot cycle super-imposed on the 11-year cycle. Several scientists consider that there is significant evidence for certain weather and climate phenomena being linked with sunspot activity.

Sunspot Index Data Centre see Federation of Astronomical and Geophysical Data Analysis Services

sunspot minima

There have been prolonged periods of very low sunspot activity from AD 1100 to 1250, from 1460 to 1550, and from 1645 to 1715. The last period is usually called the "Maunder minimum" after E. W. Maunder, the

British astronomer who, along with Gustav Sporer of Germany, first called attention to it. The Maunder minimum coincides very closely with the coldest part of the Little Ice Age.

supercooling

The cooling of liquid water to a temperature below the normal freezing point without causing the water to freeze.

surface-air temperature

The temperature of the air near the surface of the Earth, usually determined by a thermometer in an instrument shelter about 1.3m above the ground. The true daily mean, obtained from a thermograph, is approximated by the mean of 24 hourly readings and may differ by 1°C from the average based on minimum and maximum readings. The global average surface-air temperature is about 15°C.

sustainability

A term used in, for example, agriculture and forestry to designate management methods designed to ensure that the productive yield of an ecosystem is maintained undiminished for the benefit of future generations.

SWCC see **Second World Climate Conference**

synergism

The co-operative action of separate substances such that the total effect is greater than the sum of the effects of the substances acting independently.

synoptic

In general, pertaining to or affording an overall view. In meteorology, this term has become somewhat specialized in referring to the use of meteorological data obtained simultaneously over a wide area for the purpose of presenting a comprehensive and nearly instantaneous picture of the state of the atmosphere.

synoptic chart

In meteorology, any chart or map on which data and analyses are presented that describe the state of the atmosphere over a large area at a given moment in time.

synoptic observation

A meteorological observation made at the same time at many stations to obtain a general representation of the state of the atmosphere at the given time.

synoptic station

A meteorological station which makes synoptic observations at fixed times and transmits them in real time to a receiving centre.

System for Analysis, Research and Training (START)

The International Geosphere–Biosphere Programme (IGBP) has announced plans for the development of an international network of regional research centres and sites to gather data and analyze global change problems within regional contexts. This project, called the System for Analysis, Research and Training (START), is to consist of regional research networks, each network to span a scientifically coherent area, including a regional research centre with central computer and analytical facilities and many regional research sites. START will be global in conception, and high priority will be placed on securing funding for critical areas of developing countries where limited resources could prejudice the establishment of such proposed structures. The equatorial South America, Northern Africa and tropical monsoon regions have been designated as areas of highest priority.

systematic observations

Observations taken at specific places and at specific times and over a long period which may be used to "monitor" the behaviour of the atmosphere.

systems approach

A method of enquiry which complements the classical analytical method of science by emphasizing the concept of whole systems, and the irreducible properties of whole systems that result from the interactions among individual components.

T

taiga

A broad band of coniferous forest south of the Arctic tundra.

target abatement dates

Dates or schedules of compliance for significant dischargers, non-point source control measures, residual and land disposal controls, including major interim and final completion dates, and requirements that are necessary to assure an adequate tracking of the progress towards compliance.

target schedules

The time period in which commitments to achieve emission control targets of greenhouse gas emissions are to be realized.

TCP see **Tropical Cyclone Programme**

Technical Commissions of WMO

The main purpose of each of the eight technical commissions of WMO is to study and make recommendations to the WMO Congress and the WMO Executive Council on subjects within their terms of reference and on matters directly referred to the commission by Congress and the Executive Council. The members of commissions are technical experts in the fields covered by the terms of reference of the commission designated by WMO Members. A Member may designate such number of experts to serve on a commission as it deems necessary.

The WMO Technical Commissions established by the WMO Congress are classified in two groups as follows :

I. *Basic Commissions for*
   Basic Systems (CBS)
   Instruments and Methods of Observation (CIMO)
   Hydrology (CHy)
   Atmospheric Sciences (CAS)
II. *Applications Commissions for*
   Aeronautical Meteorology (CAeM)
   Agricultural Meteorology (CAgM)
   Marine Meteorology (CMM)
   Climatology (CCl)

Technical Co-operation Programme of WMO

The Technical Co-operation Programme of WMO comprises the mainstream of organized transfer of meteorological and hydrological knowledge and proven methodology among WMO Members. Particular emphasis is laid upon the development of a wide range of services related to weather prediction, climatology and hydrology; on the development and operation of key World Weather Watch infrastructures; and on supporting the Education and Training Programme of WMO. The Programme is funded mainly by UNDP, by WMO's own Voluntary Co-operation Programme (VCP) and the WMO regular budget.

tephigram

A thermodynamic diagram with rectangular cartesian or oblique co-ordinates of temperature, and potential temperature.

thermal expansion of the oceans

At constant mass, the volume of the oceans, and thus sea level, will vary with changes in the density of sea water. Density is inversely related to temperature; thus, as the oceans warm, density decreases and the oceans expand. Marked regional variations in sea-water density and volume can also result from changes in salinity, but this effect is relatively minor at the global scale. In order to estimate oceanic expansion (past or future), changes in the interior temperature, salinity and density of the oceans have to be considered, either empirically or by models. Unfortunately, such observational data are scant, in both time and space.

thermal wind

The thermal wind in a specified atmospheric layer, at a given time and place, is the vertical geostrophic windshear in the layer concerned. The term "thermal wind" was adopted because wind shear is determined by the distribution of mean temperature in the layer concerned.

thermocline

A transition layer of water in the ocean, with a steeper vertical temperature gradient than that found in the layers of ocean above and below. The permanent thermocline separates the warm mixed surface layer of the ocean from the cold deep ocean water, and is found between 100m and 1000m depths. The thermocline first appears at the 55–60°N and S latitudes, where it forms a horizontal separation between temperate and polar waters. The thermocline reaches its maximum depth at mid-latitudes, and is shallowest at the Equator and at its northern and southern limits. The transfer of water and carbon dioxide across the thermocline zone occurs very slowly. The thermocline therefore acts as a barrier to the downward mixing of carbon dioxide.

thermodynamic diagram

A diagram used for the representation of the thermodynamic state of a portion of the atmosphere.

thermogram

The record made by a thermograph.

thermograph

A thermometer used to give a graphical record of the time variations of temperature.

thermohygrograph

An instrument resulting from the combination of a thermograph and a hygrograph, so as to provide on the same diagram simultaneous time recordings of atmospheric temperature and humidity.

thermometer

An instrument used in the measurement of temperature.

Third World

The term is commonly used to designate the developing countries.

Third World Academy of Sciences (TWAS)

The Third World Academy of Sciences was founded in 1983. The specific aims of the Academy are: to help in providing high-level scientists in developing countries with the conditions necessary for the advancement of their work; to promote mutual contacts of individual research workers in developing countries among themselves and with the world scientific community; to help in developing high-level scientific expertise in developing countries by identifying young talented scientists through a recognition of their merits and by promoting the growth of their creativity; to identify persons of outstanding talents in developing countries who can advise on national and international research policies; and to encourage scientific research on major problems of the Third World and the exchange of expertise among developing countries.

thunder

A sharp or rumbling sound which accompanies lightning.

thunderstorm

A local storm, usually produced by a cumulo-nimbus cloud, and accompanied by thunder and lightning.

tidal power

Tidal power is mechanical power generated by the rise and fall of ocean tides which may in turn be converted to electrical power. The possibilities of utilizing tidal power have been studied for many generations, but the only feasible schemes devised so far are based on the use of one or more tidal basins, separated from the sea by dams or barrages, and of hydraulic turbines through which water passes on its way between the basins and the sea. The disadvantages of tidal power generation are the very high capital cost of the dams or barrages. The limitation imposed by the tidal cycle can be overcome by generating energy from both the filling and emptying of the basins to provide base-load power, and by using base-load power at times of low demand to pump water to higher reservoirs, from which it can be released to meet peak load demands.

timing of responses to climate change

The best estimates of global mean temperatures over the past 100 years suggest that overall they have risen about 0.5°C, and that sea level has risen during the same period by about 12cm. It is further considered that the current atmospheric levels of greenhouse gas emissions may already have committed the world to an additional 0.5°C of warming, and an additional 10–30cm of sea-level rise over the next fifty years, even if the atmospheric composition were stabilized immediately. An additional factor of considerable concern is that our understanding of the atmosphere and the oceans and natural ecosystems is limited, and the possibility exists that large and abrupt changes may occur which could overwhelm our adaptive capabilities, and far exceed the natural rates of change in ecosystems. The likelihood of such a climatic surprise increases as climate deviates from historical bounds.

The opinion of most scientists is that it would be inappropriate to postpone action on the greenhouse gas and climate question until the consequences of warming, which lag behind greenhouse gas emissions, are clearly visible. Policy actions already implemented, or under consideration, could limit or mitigate the consequences of warming. The timing of these initiatives to limit and adapt to climate change is critical from both an environmental and a financial perspective.

TOGA see **Tropical Ocean and Global Atmosphere Programme**

TOGA Coupled Ocean–Atmosphere Response Experiment

A major regional field experiment, similar in scope to the GARP Atlantic Tropical Experiment (GATE), is being organized to take place in 1992, within the region 140° to 180° east of the western tropical Pacific Ocean, to study the intense meso-scale and large-scale interactions between the warmest oceanic waters on Earth and the overlying atmosphere. In this connection, crucial oceanographic satellite missions, the European microwave remote-sensing satellite ERS-1, and the US–French altimetric satellite TOPEX/POSEIDON, are being readied for launch in 1991 and 1992 respectively, and will be complemented by Japan's ADEOS satellite mission in 1995. Preparations are being made to deploy large numbers of several different kinds of automatic instrumented oceanographic buoys or fixed-depth floats. Other more conventional oceanographic devices, especially moored arrays of current-meters and coastal tide-gauge stations, are in place.

tornado

A violently rotating column of air that is usually visible as a funnel cloud hanging from dark thunderstorm or cumulo-nimbus clouds. It is one of the least extensive but most destructive of all storms.

Toronto Conference (1988)

The 1988 Toronto Conference on "The Changing Atmosphere" was organized by the Canadian Government as an initial Canadian response to the Brundtland Commission's call for action on climate change. It included some 300 scientists and policy-makers from 46 countries and international organizations. Its stated objectives were to improve awareness among senior policy-makers of the importance of chemical changes in the atmosphere, and to improve the global community's ability to recognize, and avoid or mitigate critical, human-induced atmospheric changes and, in particular, their effects on society, natural resources and national economies.

The conference involved several working groups which worked in parallel to produce detailed assessments and recommendations with respect to: energy, food security, urbanization and settlement, water resources, land resources, marine resources, forecasting and futures, decision-making and uncertainty, industry, investment and trade, geopolitical issues, legal dimensions and integrated programmes.

On the basis of comments on a report presented to the final plenary session, the Conference issued a statement which concluded that:

Humanity is conducting an uncontrolled, globally pervasive experiment whose ultimate consequences could be second only to a global nuclear war. The Earth's atmosphere is being changed at an unprecedented rate by pollutants resulting from human activities, inefficient and wasteful fossil-fuel use and the effects of rapid population growth in many regions. These changes represent a major threat to international security and are already having harmful consequences over many parts of the globe.

Far-reaching impacts will be caused by global warming and sea-level rise, which are becoming increasingly evident as a result of continued growth in atmospheric concentrations of carbon dioxide and other greenhouse gases. Other major impacts are occurring from ozone-layer depletion resulting in increased damage from ultraviolet radiation. The best predictions available indicate potentially severe economic and social dislocation for present and future generations, which will worsen international tensions and increase the risk of conflicts among and within nations. It is imperative to act now.

The Conference also called upon governments, the United Nations and its specialized agencies, industry, educational institutions, non-governmental organizations and individuals, to take specific actions to reduce the impending crisis caused by pollution of the atmosphere. It asserted that no country can tackle this problem in isolation and that international co-operation is essential in the management and monitoring of, and research on, this shared resource.

The Conference Statement further called upon governments to work urgently towards an Action Plan for the Protection of the Atmosphere. It envisaged that this would include an international framework convention, as well as national legislation, to provide for protection of the global atmosphere. The Statement also called upon governments to establish a World Atmosphere Fund financed in part by a levy on the fossil-fuel consumption of industrialized countries to mobilize a substantial part of the resources needed for these measures. See also: Toronto Conference (1988)

The conference produced a consensus "call for action" including:

* general acceptance and ratification of the Montreal Protocol on Substances that Deplete the Ozone Layer;
* adoption of energy policies to reduce carbon dioxide emissions by

20% of 1988 levels by the year 2005 and by 50% as soon as
practicable;
* adoption of a target of 10% improvement in energy efficiency by 2005;
* development of a comprehensive global convention on the protection
  of the atmosphere, backed by a World Atmosphere Fund (financed in
  part by a tax on fossil fuel consumption in industrialized countries);
  and
* measures to promote intergovernmental co-operation and public
  awareness.

TOVS
A television and infrared observation satellite operational vertical
sounder.

trade winds
Persistent winds, mainly in the lower atmosphere, which blow over vast
regions from subtropical anticyclones towards equatorial regions. The
pre-dominant directions of the trade winds are from the northeast in the
Northern Hemisphere and the southeast in the Southern Hemisphere.

tradeable emission permits
Emission permits could be issued by an international authority to
individual countries who could then trade or exchange such permits
with other countries. That is, country A could for example reduce its
carbon dioxide emissions to a level below that required by any treaty,
but at the same time allow – through a transfer of its "rights to pollute"
– another country the right *not* to meet the standards set by any treaty,
provided that the net rate of emissions in the two countries was within
the protocol requirements.

A system for controlling total aggregate emissions of a pollutant such
as a greenhouse gas from a collective community could be developed by
allocating permits to community members for the right to release
quantities of the pollutant. Permits can then be traded within the
community as a marketable commodity. The community can be a specific
economic sector, a region, a country, or a group of countries. Limiting
the quantity of total allowable emissions gives a tradeable permit an
economic and marketable value. Such trading helps a community to
meet its collective emission-control targets at a least net cost.

tree line
A line beyond which climatic conditions do not support the growth of
high-wooded vegetation. This limit may be topographical (height) or
geographical (latitude).

tree rings
Concentric rings of secondary wood evident in a cross section of the
stem of a woody plant. The difference between the dense, small-celled

late wood of one season and the wide-celled early wood of the next enables the age of a tree to be estimated.

tropical cyclone see **hurricane**

Tropical Cyclone Programme (TCP)

The main purpose of the Tropical Cyclone Programme (TCP) of WMO is to assist WMO Members, through an internationally co-ordinated programme, in their efforts to mitigate tropical cyclone disasters. The programme is effected on both national and regional levels through co-operative action. It covers activities of WMO Members, WMO regional associations, other international and regional bodies, and the WMO Secretariat. The main long-term objective of the programme is to minimize the loss of life and damage from tropical cyclones and associated phenomena.

There is, unquestionably, growing potential for massive disaster in areas prone to tropical cyclones. Increasing populations and continuing physical developments in disaster-prone coastal areas provide the elements for calamities on a scale seldom seen hitherto. Humankind's best efforts will be needed to offset these increasing risks. Concerted efforts must be made to harness both current knowledge and the potential deriving from advances in science and technology for the welfare of populations in tropical cyclone areas. This situation certainly presents expanding opportunities and challenges to be addressed by the members concerned, with an important rôle to be played by the TCP.

One of the major problems faced by many meteorological services in the developing world is that they are unable to command an adequate status in the national hierarchy. Meteorology is frequently not seen by senior government officials as being vital to national interests. This problem is compounded by the fact that, in many cases, convincing arguments, backed up by figures to demonstrate the contribution meteorological services can make to the national prosperity, have not been well elaborated and submitted to the relevant authorities.

Efforts to reduce the impact of natural hazards such as tropical cyclones suffer additionally from the fact that their irregularity and the uncertainty of their future occurrences foster the notion that they do not qualify for high-priority urgent attention. The argument that resources should be devoted to cases where the benefits are not assured within a defined time frame is difficult to rebut, particularly in cyclone-prone areas which have not recently experienced severe cyclones. Yet this situation is precisely the background to many of the major cyclone disasters of the past.

Two specific projects in the period 1992–2001 should be noted:
Project 18.3 – Tropical cyclone and storm-surge simulation, forecasting and warning. This project will assist WMO Members in upgrading tropical cyclone and storm surge forecasting and warning capabilities through technical aspects, and co-ordination and co-operation.

Achievement of this project will lead to improvement in the operational forecasting of changes in intensity and movement of tropical cyclones including forecasting of landfall, in short-range forecasts and warnings including precipitation forecasting, and in forecasting and warning systems for storm surges associated with tropical cyclones.

Project 18.5 – Development of tropical cyclone mitigation systems and promotion of public information. This project will assist WMO Members in ensuring the wide dissemination and the effectiveness and appropriate response to tropical cyclone warnings in close co-operation with ESCAP, UNDRO, LRCSS and other bodies with special expertise in the respective fields. Achievement of this project will encourage the establishment of a natural disaster mitigation system for tropical cyclones with increasing understanding of the threat and impact of tropical cyclones. It will also lead to protective measures being taken.

Tropical Ocean and Global Atmosphere Programme (TOGA)

The Tropical Ocean and Global Atmosphere (TOGA) study is a ten-year programme, led by the WMO/ICSU Joint Scientific Committee for the WCRP, in co-operation with ICSU's Scientific Committee on Oceanic Research, the UNESCO Intergovernmental Oceanographic Commission, and their joint Committee on Climate Changes and the Ocean.

The objective of TOGA is to achieve a description of the large-scale transient variations of tropical ocean basins and the global atmosphere, to determine the extent to which this system is predictable on timespans of months to several years, to understand the mechanisms underlying this predictability, and to develop the means (both field observations and coupled atmospheric–oceanic models) to achieve effective predictions. The main scientific targets of the TOGA programme are the El Niño/Southern Oscillation phenomenon and the variations in the monsoon regime from month to month and year to year.

TOGA has initiated a range of specific projects to establish or complement oceanic and atmospheric observing systems in the tropical zone and the Southern Hemisphere. These projects include launching temperature probes from merchant vessels to provide systematic bathy-thermographic sections every month or every second month along selected shipping routes across the three tropical ocean basins, surface pressure and temperature measurements from automatic drifting buoys in the Southern Hemisphere, real-time acquisition and transmission of tide-gauge data, and upper-air wind observations at several island stations along the Equator. Data from TOGA systems and existing operational networks already serve as a basis for experimental predictions of El Niño or similar events in the tropical Pacific, using coupled atmosphere–ocean models. Other global modelling studies are being pursued to investigate, in particular, the sensitivity of the global atmosphere to localized anomalies in sea-surface temperature and the variability of monsoonal regimes.

Tropical Urban Climate Experiment (TRUCE)

The formulation of a proposal for a "tropical urban climate experiment" was first discussed at a meeting in Mexico City in 1984. The Tenth Session of the WMO Commission for Climatology elaborated this further, and a plan of action was approved by the Executive Council of WMO in 1989.

The purpose of TRUCE is to improve our understanding of the controlling mechanisms associated with the modification of the climate in tropical urban areas. It will also provide a better scientific basis for decisions relating to urban planning and environmental measures, as well as the rôle of operational meteorological services.

The long-term objectives of TRUCE are:
* to produce a global database of structured data and the characteristics of various tropical urban climates;
* to initiate, co-ordinate and implement observational and theoretical research programmes;
* to develop models capable of simulating the urban climate system on different time- and space scales;
* to arrange for the results of TRUCE-related activities to be made available in the form of practical guidelines, thereby promoting the use of climatological information in building and urban design; and
* to forge links between researchers.

tropopause

The boundary between the troposphere and the stratosphere. It represents the point at which temperatures stop falling prior to the isothermal conditions of the lower stratosphere. The tropopause is not a continuous surface between the tropics and poles, but is broken at the latitude of the subtropical jet stream and is suddenly lower on the poleward side. These breaks enable mixing to take place between tropospheric and stratospheric air, which would otherwise not occur because the isothermal layer of the stratosphere acts as a stable inversion.

troposphere

The lowest layer of the atmosphere, where almost all weather phenomena develop. It takes its name from the Greek word tropos, meaning "a turn": it is the atmospheric layer where turning and convective mixing is dominant.

TRUCE see **Tropical Urban Climate Experiment**

tundra

A type of ecosystem dominated by lichens, mosses, grasses, and woody plants. It is found at high latitudes (arctic tundra) and high altitudes (alpine tundra). Arctic tundra is underlain by permafrost and is usually very wet.

turbidity

The reduced transparency of the atmosphere, caused by absorption and scattering of radiation by solid or liquid particles, other than clouds, held there in suspension.

turnover rate

The fraction of the total amount of mass (e.g. carbon) in a given pool or reservoir that is released from or enters the pool in a given length of time. Turnover rate of carbon is expressed as Gt C/year.

TWAS see **Third World Academy of Sciences**

typhoon see **hurricane**

Typhoon Committee

An intergovernmental committee responsible for promoting and co-ordinating the planning and implementation of measures required for minimizing typhoon damage in the ESCAP region.

U

ultraviolet radiation

Electromagnetic energy of higher frequencies and shorter wavelengths than visible light. Ultraviolet radiation is divided into three ranges: UV-A (320–400nm), UV-B (280–320nm) and UV-C (40–290nm).

UNCED see **United Nations Conference on Environment and Development – 1992**

UNCTAD see **United Nations Conference on Trade and Development**

UNCTC

The United Nations Centre on Translational Corporations.

UNDRO

The United Nations Disaster Relief Co-ordinator.

UNECE

The United Nations Economic Commission for Europe.

UNEP see **United Nations Environment Programme**

UNESCO see **United Nations Educational, Scientific and Cultural Organization**

UNIDO
The United Nations Industrial Development Organization.

UNITAR
The United Nations Institute for Training and Research.

United Nations Conference on Desertification (1977)

A UN Conference on Desertification was held in Nairobi, Kenya, in 1977, and was attended by representatives of 95 countries and over 60 international organizations, both governmental and non-governmental.

In advance of the conference, studies were carried out with the purpose of providing guidelines as to the decisions and recommendations that might be reached by the conference. The main studies consisted of:
* case histories of areas where desertification was a problem;
* background documents on climate and desertification, ecological change and desertification, population, society and desertification, and technology and desertification;
* a world map of desertification prepared by FAO with the collaboration of UNESCO and WMO.

It was very clear from the results of the studies that, whereas in most cases marginal climatic conditions were conducive to desertification, climate change as such was not the root cause of the problem. A primary cause was shown to be over-exploitation of the land, whether agricultural or pastoral, in circumstances where the ecological balance was highly sensitive. Another cause was shown to be lack of overall planning regarding land-use. The studies identified 24 countries where all or nearly all the land not already desert was at risk and 23 countries where much of the land was at risk.

The conference adopted a plan of action with a target date of the year 2000 for implementation. Subjects covered included:
* the impact of recurring droughts and climate fluctuation on land management;
* the need for adequate networks of meteorological, climatological and hydrological stations in areas of concern;
* monitoring of atmospheric changes, dust transport and changes in irrigated lands;
* assessment of water needs and the reduction in water losses by evaporation;
* building climatology and associated information for human settlements.

United Nations Conference on Environment and Development: 1992 (UNCED)

Strategies for dealing with issues arising from socio-economic development processes were the focus of the world's largest international conference on the environment, held in Rio de Janeiro,

Brazil, from 1–12 June 1992. Many states were represented at the level of head of state or government, and arrangements for the broad participation of non-governmental organizations (especially from developing countries) were made. In calling for this conference the United Nations General Assembly, in its Resolution 44/228 dated 22 December 1989, stressed the need to find integrated strategies that would prevent further degradation of the environment and promote sustainable, environmentally sound development in every part of the world. The issues addressed by the conference were organized in six principal components:

* an "Earth Charter" or declaration of basic principles for the conduct of nations and peoples in respect of environment and development;
* agreements on specific legal measures, e.g. conventions for the protection of the atmosphere and biological diversity which were negotiated before the Conference and signed or agreed to at the conference;
* an agenda for action, "Agenda 21", establishing the agreed work-programme of the international community for the period beyond 1992 and into the 21st century with respect to the issues to be addressed by the conference, with priorities, targets, cost estimates, modalities and assignment of responsibilities; plus the means to implement this agenda, including new and additional financial resources, transfer of technology, and strengthening of institutional capacities and processes.

Several meetings of the Preparatory Committee were set up to oversee preparations for the 1992 conference. At its organizational meeting in New York in March 1990, it created two working groups to study air, land and water resources. Working Group I deals with protection of the atmosphere, protection and management of land resources, conservation of biological diversity, and environmentally sound biotechnology. Working Group II concentrates on protection of the oceans, seas and coastal areas, protection of the quality and supply of freshwater resources, and environmentally sound management of toxic chemicals and wastes. A third Working Group was later established to deal with legal, institutional and related matters.

Specifically, the conference is expected to have produced:

* an Earth Charter that will embody basic principles which must govern the economic and environmental behaviour of peoples and nations to ensure "our common future";
* Agenda 21, a blueprint for action in all major areas affecting the relationship between the environment and the economy; it will focus on the period up to the year 2000 and extend into the 21st century;
* the means to carry out the agenda by making available to developing countries the additional financial resources and environmentally sound technologies they require to participate fully in global environmental co-operation, and to integrate environmental considerations into development policies and practices;

* agreement on strengthening institutions in order to implement these measures;
* conventions on climate change, biological diversity and, perhaps, forestry may be negotiated prior to the conference and signed or agreed to in Brazil.

The Preparatory Committee, open to all member states, held its first substantive session in August 1990 in Nairobi, Kenya. In 1991, the Preparatory Committee and the Working Groups convened in Geneva from 18 March to 5 April and from 12 August to 4 September. The final session took place in New York in February/March 1992. Through its three Working Groups, the Committee is dealing with several substantive areas, including protection of:

* the atmosphere (climate change, ozone layer, air pollution;
* land resources;
* freshwater resources;
* oceans, seas and coastal areas.

UNCED (the Earth Summit) will be a conference of leaders of governments, but its success largely depends on the interest and support of peoples and their active participation in NGOs and citizens groups which contributed to its preparations. Provision was made for relevant NGOs to participate in meetings of the Preparatory Committee. A parallel event was held by non-governmental groups in Rio de Janeiro at the time of UNCED.

United Nations Conference on the Human Environment (1972)

The conference, held in Stockholm in 1972 – among other things – approved the establishment of the United Nations Environment Programme (UNEP). The conference adopted 109 recommendations and the importance of meteorology was reflected in no fewer than 35 of them. These recommendations and other matters arising from the conference were considered by the General Assembly of the United Nations in December 1972. A Governing Council for UNEP was established and at its first meeting (Geneva, June 1973) it laid down a series of programme areas on which action was to be initiated without delay. WMO had already begun discussions with UNEP, and projects were soon developed on the background monitoring of air pollution, the physical basis of climate, marine pollution, the forecasting of drought, the development of water resources, and desertification. Many ideas emerged from the conference and were pursued in a series of UN World Conferences during the ensuing few years covering such subjects as: food (Rome, 1974), population (Bucharest, 1974), women (Mexico City, 1975), human settlements [Habitat] (Vancouver, 1976), water (Mar del Plata, Argentina, 1977), and desertification (Nairobi, 1977).

United Nations Conference on Human Settlements (1976)

In 1976, a sequel to the Stockholm Conference on the Environment took place in Vancouver, Canada, in the form of a World Conference on Human Settlements, generally referred to as the HABITAT Conference. Among the main reasons for the conference were the need to prevent further degradation of the environment and the urgency to provide dwellings to millions living under deplorable conditions in consequence of catastrophic population imbalances. The preparatory work had concentrated on such pressing issues as population growth, migration from rural to urban areas, inadequate control over land-use, social injustices, and the need for planned urban development. The HABITAT meeting was one of the largest conferences ever convened by the United Nations. It was attended by over 1,000 delegates, representatives and observers from 131 countries, 29 UN and other world and regional agencies, and 142 non-governmental organizations.

United Nations Conference on Trade and Development (UNCTAD)

A principal UN forum in which developed and developing countries meet to discuss matters relating to trade and development.

United Nations Educational, Scientific and Cultural Organization (UNESCO)

UNESCO is a United Nations agency, founded in 1945, to support and complement the efforts of member states to promote education, scientific research and information, and the arts, and to develop the cultural aspects of world relations. It is financed by its own budget and also draws on funds pledged by member states to the UN Development Programme. Its headquarters are in Paris.

UNESCO is involved in climate issues, particularly through the International Oceanographic Commission (IOC). The IOC is a collaborating partner in the World Climate Programme and has several specific areas of interest including:
* co-ordination of the Global Sea Level Monitoring System (GLOSS) which monitors and analyzes sea-level data,
* work on marine science and ocean services for development, together with a major assistance programme to enhance the marine science capabilities of developing countries.

United Nations Environment Programme (UNEP)

A UN programme charged with the co-ordination of intergovernmental measures for environmental monitoring and protection. Formed after the 1972 UN Human Environment Conference. UNEP's first Executive Director was Mr Maurice Strong, who had been Secretary General of the 1972 conference. He was succeeded in 1976 by Dr Mostafa K. Tolba. The Programme is operated through a secretariat located in Nairobi.

UNEP is actively involved in climate change issues – among other things – through its participation in the IPCC, the World Climate Programme, and the AGGG. It has also been directly involved as the lead agency in

the implementation of the Vienna Convention and then the Montreal Protocol on Substances that Deplete the Ozone Layer.
(See also **Global Environment Monitoring System**, and **World Climate Impact Assessment and Response Strategies Programme**)

United Nations General Assembly Resolution 43/53 (1988)

The 1988 UN Resolution 43/53 on the "Protection of Global Climate for Present and Future Generations of Mankind" urged ". . . governments, intergovernmental and non-governmental organizations and scientific institutions to treat climate change as a priority issue, to undertake and promote specific co-operative active-oriented programmes and research so as to increase understanding on all sources and causes of climate change . . ."

United Nations General Assembly Resolution 44/207 (1989)

The 1989 UN General Assembly Resolution 44/207 recommends that governments, with due consideration of the need for increased scientific knowledge of the sources, causes and impacts of climate change and of global, regional and local climates, continue and, wherever possible, increase their activities in support of the World Climate Programme and the International Geosphere–Biosphere Programme, including the monitoring of atmospheric composition and climate conditions. It further recommends that the international community support efforts by developing countries to participate in these scientific activities.

V

vapour pressure

The pressure exerted by water vapour in the atmosphere.

varve

A layer of sediment deposited in lakes during one year. Each layer consists of two parts, deposited at different seasons and differing in colour and texture; thus, the layers can be counted and measured. In a complete series, the number of layers gives the date on which the ground was vacated by the retreating ice.

VCP See **Voluntary Co-operation Programme (WMO)**

Vienna Convention

The Vienna Convention for the Protection of the Ozone Layer of 22 March 1985 which was entered into force in August 1988.

Villach Conference 1985

A major scientific assessment of "the rôle of carbon dioxide and of other greenhouse gases in climate variations and associated impacts" was carried out by the scientific community in the 1983–85 period and was presented at a conference held in Villach (Austria) in October 1985, organized by WMO, UNEP and ICSU. The Conference, which was attended by 100 experts from 30 countries, suggested that ". . . the problem of a possibly changing climate due to the emissions of greenhouse gases should be considered as one of today's most important long-term environmental problems", and that ". . . the understanding of the greenhouse question though still incomplete, is sufficiently developed that scientists and policy-makers should begin an active collaboration to explore the effectiveness of alternative policies and adjustments". The Conference concluded ". . . While some warming of climate now appears inevitable due to past actions, the rate and degree of future warming could be profoundly affected by governmental policies on energy conservation, use of fossil fuels, and the emission of some greenhouse gases".

Among the principal findings of the Villach 1985 Conference were:

* Many important economic and social decisions are being made today on long-term projects such as irrigation and hydropower, drought relief, agricultural land-use, structural designs and coastal engineering projects, and energy planning – all based on the assumption that past climatic data are a reliable guide to the future. This is no longer a good assumption since the increasing concentrations of greenhouse gases are expected to cause a significant warming of the global climate in the next century.

* The rôle of greenhouse gases other than CO_2 in changing the climate is already about as important as that of CO_2 itself. If present trends continue, the combined concentrations of all of the greenhouse gases would be equivalent to a doubling of the CO_2 concentrations from pre-industrial levels, possibly as early as the 2030s.

* While other factors such as aerosol concentrations, changes in solar-energy input, and changes in vegetation may also influence climate, the greenhouse gases are likely to be the most important cause of climatic change over the next century.

* There is little doubt that future changes in climate of the order of magnitude obtained from climate models for a doubling of the atmospheric CO_2, concentration could have profound effects on global ecosystems, agriculture, water resources and sea ice.

* Governments and regional and intergovernmental organizations should take the results of the Villach 1985 assessment into account in their policies for social and economic development, environmental programmes and policies for the control of greenhouse gas emissions.

* Work should be started immediately on the analysis of policy and economic options, when the widest possible range of social responses aimed at preventing or adapting to climate change should be identified, analyzed and evaluated.

Villach Technical Workshop 1987

The scientific consensus reached at the Villach 1985 Conference was used as the starting point for the Villach Technical Workshop on Developing Policies for Responding to Climatic Change, held in September 1987 (followed by a senior policy-makers' conference in Bellagio (Italy) in November 1987) with the following main themes: possible scenarios for future changes of climate and sea-level; effects of possible climate changes on regions in the high latitudes, middle latitudes, humid tropics, semi-arid tropics and coastal zones; management options for responding to the possible changes; and considerations that might bear on policy development.

The importance of considering the rates of change of climate, sea level, sea ice, vegetation, etc., was also considered, and the Workshop found that the concept of setting limits on the rates of change might have important implications for the choice of policy options. The Workshop also gave a clear message that scientific assessments of climate change must address three issues if they are to be useful for policy discussions: the rate and timing of climate change, the uncertainties in forecasts of climate change, and the regional distribution of climate changes.

The scientific understanding of the impact of the greenhouse effect on global temperatures as reported in the findings of the 1987 Villach Workshop shows a range of scenarios of global temperature change that might plausibly occur between now and the end of the next century. All scenarios quickly carry the world into a condition of higher temperature and sea level than has been experienced within the past 100 years. The intensity of the global hydrological cycle in terms of precipitation and evaporation is expected to increase by 2–3% for each degree of global warming. The changed global climate can therefore be expected to be wetter than that of the recent past.

The wide range of scenarios reflect uncertainties in two basic areas: future patterns of fossil-fuel use, rates of reforestation and de-forestation, and other activities leading to greenhouse gas emissions; and the response of the climate system to a given level of greenhouse gases. These uncertainties contribute about equally to the overall uncertainty in forecasting future climate change. The envelope of scenarios was constructed so that, in the judgement of the 1987 Villach Workshop, there is a 90% chance that the actual future pattern of climate change will lie within the bounds set by the "upper" and "lower" curves.

The "upper" curve (of the scenario) represents the temperature and sea-level change that could result if there is a radical expansion of fossil-fuel use and other activities that emit greenhouse gases, and if the climate's response to greenhouse gases exhibits the high sensitivity predicted by a few studies. In contrast, the "lower curve" represents the change that could result if there is a radical curtailment of fossil-fuel use and other activities that emit greenhouse gases, and if the climate's response exhibits the relatively low sensitivity predicted by a few other studies. The "middle curve" shows the temperature and sea-level change that could result if there is no change in the present rates of increase of greenhouse gas emissions, and if the climate response exhibits the moderate sensitivity to greenhouse gases that the majority of studies predict. In calculating this curve, the Villach Workshop assumed that the 1987 Montreal protocol on protecting the ozone layer will be successfully implemented, thus reducing the emissions of chlorofluorocarbons significantly below their recent rates of increase.

Uncertainties in the forecasts of regional climatic responses are even greater than those in the forecasts of global climate response, and the 1987 Villach Workshop thinking about possible regional climate changes indicates that the greatest warming is likely to occur during the winter in the high latitudes of the Northern Hemisphere, with changes 2.0 to 2.4 times greater and faster than the globally averaged annual values. In contrast, temperature changes in the low latitudes will probably be somewhat smaller and slower than the globally averaged values. Regional precipitation forecasts are the most uncertain of all. However, the studies suggest that major changes could include enhanced winter snowfall in the high latitudes, intensified rains in the presently rainy low latitudes and, perhaps, a decrease in summer rainfall in the middle latitudes.

volcanic dust

Dust ash, or other particulate matter commonly suspended in the atmosphere after volcanic eruptions. After explosive eruptions the dust may be thrown to heights of 20–30 km or more. The fall-out times of dust particles are quite short (a matter of days or weeks) depending on altitude and precipitation, but volcanogenic aerosols, usually sulphates, may linger for months, spreading as a long-lived veil in the stratosphere over much of the Earth. Massive volcanic explosions such as those of Mount St Helens in May 1980 put into the stratosphere myriads of submicroscopic-size rock particles and aerosols derived from sulphur dioxide. The volcanic matter typically passes around the Earth in ten days to a few weeks and gradually spreads into an increasingly uniform veil which may cover the whole hemisphere or even the whole Earth within about half a year. The particles are so small that they take from twenty days to a year to fall 1km, and are liable to stay in the stratosphere for one to seven years.

The effect of their partial interception of the solar radiation is to warm the dust layer while, at the Earth's surface, temperatures fall somewhat below what they would otherwise be. The cooling, at its maximum in the first year, has been assessed at from 0.1°C to about 1.0°C. On a scale which ranks volcanic dust veils in terms of the mass of material initially ejected and the duration and maximum spread of the veil, the Krakatoa eruption in Indonesia in 1883 is ranked at 1000. In 1902 a group of eruptions in the West Indies produced a new veil ranked as 1000 and this was renewed at least over the northern polar regions in 1912 by the Katmai eruption in Alaska. The next large eruption was Mt Agung in

Bali in 1963 with a dust veil index of 800. This was followed by eruptions of other volcanoes and by the late 1960s the dust veil index was again rated at over 1,000.

Voluntary Co-operation Programme of WMO

WMO's Voluntary Co-operation Programme (VCP) allows individual countries to make requests for assistance of various types, and individual donors agree to support those of the requests that they are able and willing to fund. Priorities for meteorological activities are set by both recipient and donor countries. In this case, a mechanism to oversee this programme was established by WMO's Executive Council through its VCP Panel of Experts and Informal Planning Meetings.

The Voluntary Co-operation Programme consists of two elements: VCP(F) and VCP(ES). The VCP(F) element is a multilateral fund from which the Executive Council of WMO makes allocations each year. In 1989, funds were allocated for spare parts of equipment, travel of experts, fellowships, and several high-priority VCP co-ordinated programmes. The primary focus of the VCP(F) is support for operation of the WWW Programme. The VCP(ES) element, which makes up 95% of the overall VCP, consists in the provision of equipment, experts' services, and fellowships, which are donated in response to specific requests from developing countries.

vorticity

The vorticity at a point in a fluid is a vector which is twice the local rate of rotation of a fluid element. The component of the vorticity in any direction is the circulation per unit area of the fluid in a plane normal to that direction.

vorticity equation

The vorticity equation as used in meteorology relates the rate of change of the vertical component of vorticity to the horizontal divergence.

W

wait-and-see versus risk minimization

Despite recent progress in putting global warming onto the international agenda, the debate over the greenhouse effect continues to be shaped by two diametrically opposed viewpoints. These views can be characterized as follows:
* Don't act until you are certain, or wait-and-see. Analysts holding to this view believe that current scientific uncertainties are still too large to warrant costly preventive action. Instead, more research should be pursued to reduce scientific uncertainties.

* Act now to minimize risks. Those holding to this view believe that current uncertainty cuts both ways: if major warming should come true, inaction could have catastrophic consequences. Society should therefore pursue investments and policies now to minimize such risks.

The competition between these two viewpoints revolves around the following issues:
* Which aspects of the global-warming threat are scientifically established facts, and which are not?
* How costly would prevention be compared to adaptation?
* Would there be winners and losers, or would the consequences of global warming be catastrophic for the world as a whole?
* Is there reason to believe that remaining uncertainties could be satisfactorily resolved in a timeframe that would still allow preventive global action later?
* Could improved scientific modelling tools be able to distinguish reliably between winners and losers?

Walker circulation
A zonal circulation of the atmosphere confined to equatorial regions and driven principally by the oceanic temperature gradient extending across the Pacific Ocean from Indonesia to close to the Peruvian coast, and forming a component of the Southern Oscillation. In the Pacific, air flows westwards from the colder eastern area to the warm western ocean, where it acquires warmth and moisture and subsequently rises. A return flow aloft and subsidence over the eastern ocean completes the cell.

water budget
A budget of the incoming and outgoing water from a region, including rainfall, evaporation, runoff and seepage, with special attention to evapotranspiration from vegetation.

water cycle
The movement of water between and within the atmosphere and ground surface on a global scale. It involves the balance between precipitation, evaporation, advection of moisture in the air, ocean-current circulations, and river runoff on land. There is a long-term balance, so that no area of the Earth is continuously losing or gaining moisture.

water resources
About 60% of the Earth's land surface is "supported" by freshwater resources. The remaining areas are permanently frozen or arid. Water rotates through the hydrological cycle via evaporation, precipitation and runoff, and humans manage it by developing diversion, pumping, storage, transfer, treatment and distribution facilities. Agriculture accounts for 80% of global water consumption (over 90% in some

developing countries), with approximately 15% of the world's crop land receiving some irrigation. The remaining water is consumed in industrial and domestic uses. Half of the world's population lacks access to adequate freshwater supplies. Population growth and rising standards of living will increase demands for this resource.

By affecting certain components of the hydrological cycle – especially precipitation and runoff – a change in climate can alter the spatial and temporal availability of water resources. Indeed, much of what we think of as water resources management – e.g. reservoir storage and attempts to balance water supply and demand through conservation efforts – can be seen as efforts to manage intra- and inter-annual climate variability as expressed in changes in precipitation, runoff, and evapotranspiration and user demands.

water-supply sensitivity

Climate fluctuations can lead to water-supply effects ranging from minor swings that can, nevertheless, be quite disruptive in heavily developed watersheds, to failures with long-term impacts on economic development. Climate impacts on water may also uncover or exacerbate other environmental and social effects of water development. For instance, reduced streamflows can concentrate the effects of pollutants or exacerbate the spread of water-borne disease. Climate fluctuations can also affect the success of agricultural schemes associated with water systems and may impact populations displaced by water projects.

The basic unit studied in water resources is the water-supply system, which can range from a shallow well serving a small settlement to large management systems that command the water resources of several thousand square kilometres and multiple drainage basins. While the focus here is on large, integrated water systems, climate fluctuations may acutely affect people relying on informal or small-scale water resources. We know much less about water usage outside of centrally managed supply systems, especially in the developing countries. For example, village water use is related to a complex set of source and user attributes (e.g. perceived quality and reliability) that might not be apparent to the casual observer.

Governments and impact assessors are mostly interested in large water systems. Integrated surface-water systems include facilities such as diversions, aqueducts, storage reservoirs, filtering and treatment plants, and distribution networks – all managed by a central authority or consortium of users. Integrated systems may supply water for irrigation, domestic consumption, industry (including, for example, manufacturing, mining and energy development), as well as generating hydro-power and managing water flows to maintain environmental quality.

Two types of surface-water supply systems are of interest in climate impact assessments: run-of-river operations that do not store significant amounts of water, and reservoir-based systems that smooth out intra- and inter-annual variations in runoff by storing excess water for later release. Water systems in humid areas may involve simply extracting supplies from permanent streams and rivers as needed. In some more heavily developed humid zones, where demand closely matches reliable river discharges, storage reservoirs and inter-basin transfers have been developed to minimize the impacts of dry spells, as well as to provide flood control, hydro-power, navigation and recreation.

water vapour

Water vapour, although the most powerful of all the greenhouse gases, is not itself a primary factor in the enhancement of the greenhouse effect. It will, however, be a factor in both positive and negative feedback effects as temperatures begin to rise. Since warmer air can hold more moisture, and higher temperatures will cause more water to evaporate from the Earth's surface, the water vapour content of the atmosphere will increase. This will add to the greenhouse effect, but the additional moisture in the atmosphere will also increase cloud formation. The effect of this will be to reduce the amount of solar energy warming the Earth's surface, thus partly offsetting the increase in greenhouse warming caused by the water vapour.

water vapour feedback

A positive feedback (interaction) process in which an increase in the amount of water vapour increases the absorption of longwave radiation, thereby contributing to a warming of the atmosphere. Warming, in turn, may result in increased evaporation and an increase in the initial water vapour anomaly. This feedback, along with carbon dioxide, is responsible for the greenhouse effect, and operates virtually continuously in the atmosphere.

WCAP see **World Climate Applications Programme**

WCASP see **World Climate Applications and Services Programme**

WCDMP see **World Climate Data and Monitoring Programme**

WCDP see **World Climate Data Programme**

WCIP see **World Climate Impact Studies Programme**

WCIRP see **World Climate Impact Assessment and Response Strategies Programme**

WCP see **World Climate Programme**

WCRP see **World Climate Research Programme**

weather lore

Empirical weather-forecasting rules, worldwide in origin, many of which are expressed in rhyme. They include rules based on the influence of the Moon and tides, the appearance of plants and trees, the behaviour of animals, the weather prevailing on specified key dates, and the colour and appearance of the sky.

weighted values

The values of quantities to which have been attached a series of weights in order to make proper allowance for their relative importance.

West Antarctic ice sheet

Most of the early attention to the issue of sea-level rise and greenhouse warming was related to the stability of the West Antarctic ice sheet. Parts of this ice sheet are grounded far below sea level and may be very sensitive to small changes in sea level or melting rates at the base of adjacent ice shelves. In case of a climatic warming, such melting rates could increase and lead to the disappearance of ice rises (places where the floating ice shelf runs aground). Reduced back stress on the main ice sheet and larger ice velocities may result, with subsequent thinning of the grounded ice and grounding-line retreat. It is hard to make quantitative statements about this mechanism, but there is no firm evidence to suggest that the Antarctic ice sheet in general, or the West Antarctic ice sheet in particular, have contributed either positively or negatively to past sea-level rise. In fact, on the whole, the sensitivity of Antarctica to climatic change is such that a future warming should lead to increased snow/ice accumulation and thus a negative contribution to sea-level change.

westerlies

A zone, lying between the approximate latitudes 35° and 65° in both the Northern and Southern Hemispheres, in which the air motion is mainly from west to east, especially in the high troposphere and low stratosphere.

WHO

The World Health Organization.

wilting point

The point at which the soil contains so little water that it is unable to supply water at a rate sufficient to prevent permanent wilting of plants.

wind chill

A measure of the chilling effect experienced by the human body when strong winds are combined with freezing temperatures. The higher the wind chill, the faster the rate of cooling. The wind-chill factor is expressed in watts per square metre or in °C as an equivalent temperature.

wind direction
The direction *from* which the wind blows.

Winners and Losers in the Context of Global Warming: report of a workshop held in Malta, June 1990
In recent years, the issue of climate change has rapidly advanced to the top of national and international scientific agendas. As this issue has gained in importance, scientists and policy-makers have commented on the advantages and disadvantages that might accrue to nations and regions if the climate changes in coming decades. These comments have varied from the extreme – that everyone will win or that everyone will lose – to suggestions that certain nations, subnational regions, or economic sectors, may derive relative advantages or disadvantages. Such generalizations about gains and losses, advantages and disadvantages, have in most if not all cases not been based on adequate scientific assessments of possible costs/benefits or advantages/disadvantages over varying timescales.

In a first formal attempt to address the methodological approaches and constraints on their utility for the objective assessment of advantages and disadvantages associated with a climate change, a workshop "On assessing winners and losers in the context of global warming" was held in Malta from 18–21 June 1990. The workshop was organized by the Environmental and Societal Impacts Group (ESIG) of the National Center for Atmospheric Research (NCAR), with support from the United Nations Environment Programme (UNEP) and the US National Science Foundation (NSF). The overriding objective of the workshop was to stimulate discussion of the methods that might be used for, and of the constraints on, making objective assessments of the societal impacts that may derive from climate change at all levels of social organization from the local to the international levels.

The main conclusions of the Workshop were as follows:
* More research must be undertaken to develop objective method-ologies for the assessment of winners and losers in the context of climate change, considering in particular the difficulties in measurement and aggregation across varying scales of space and time.
* To improve assessment methodology, detailed case studies should be undertaken to test and evaluate different methods. Priority should be given to conducting these studies in regions where climate is currently perceived as a limiting factor in socio-economic devel-opment. In addition, criteria should be applied to case studies of the regional impacts of today's global climate regime in order to determine advantages and disadvantages. Once competing methods have been calibrated for their levels of objectivity and accuracy, they

can be considered for assessment of gains and losses resulting from a global climate change.

* There is a need for reliable and credible regional studies of the societal impacts of climate change and responses to it. Difficulties in undertaking such studies will be exacerbated by the different ways in which scientists of different backgrounds filter information and by the compounding of uncertainty along the chain of events from emissions of greenhouse gases to climate change, to impacts on ecosystems and on societies. The last of these feeds back to changes in emissions.

* The perceptions of policy-makers govern their decisions. These perceptions are based on their experience, interests, and evaluations of how their specific constituencies may be affected. In particular, it is often difficult to secure political leadership on issues such as climate change, where important constituencies may be held accountable and thus forced to make significant sacrifices.

* In the search for appropriate responses to the impacts of climate change, those countries, regions, sectors and populations that can present convincing rationales for actions from which they are likely to benefit are likely to be advantaged. This suggests that some possible response strategies may not be thoroughly examined in spite of their potential value.

* Those involved in communication and education must portray assessments of the advantages and disadvantages deriving from climate change in ways that identify their degree of certainty, reach the public, lead to accurate perceptions of climate variability and change, and consider how potential responses relate to issues of equity and values.

* The possibility of anthropogenic climate change raises important issues of international and intra- and intergenerational justice.

* Certain actions can be taken now by all countries to assess and address existing inequities that would help disadvantaged countries, subnational regions, economic sectors, and populations to respond to climate change.

* The "polluter pays" principle, including compensation for harm, is a useful starting point for determining responsibility for anthropogenic climate change.

* To minimize the adverse effects on societies of climate change and determine how the burdens of preventive actions should be shared requires the following: co-operative actions involving nations, intergovernmental institutions, and non-governmental institutions; the evolution of the international legal system; democratization; local participation in the planning and implementation of policies; and improved assessment methodologies.

* Global concern about climate change should also be translated into global responsibility for the sustainable development of the developing countries through international financial, technical and scientific co-operation.

In addition the following general recommendations were highlighted by the Workshop:
* There is a need for objective, reliable assessments of how nations, sectors, regions, and populations might be advantaged or disadvantaged with climate change.
* There is a need for improved research on the perceptual aspects of the global warming issue, including the rôle of the media in forming climate change perceptions, and the rôle of the perceptions of political leaders and how they affect the policy process on this issue.
* Attention should be focused on issues of climate change and intra- and intergenerational equity issues.
* Case studies should be undertaken at the regional level to develop methods for assessing gains and losses that might accompany a climate change.

WIPO
The World Intellectual Properties Organization.

WMO see **World Meteorological Organization**

WMO Satellite Activities (SAT)
The main purpose of the WMO Satellite Activities is to co-ordinate environmental satellite matters and activities throughout all the WMO programmes and to give guidance to the WMO Secretariat, Technical Commissions and Regional Associations on the potentialities of remote-sensing techniques in meteorology, hydrology, related disciplines and their applications. The broad utilization of satellite data within all WMO programmes requires that the Satellite Activities oversee and co-ordinate with many programmes. There is therefore direct co-ordination with the following WMO programmes: World Weather Watch, World Climate, Research and Development, Hydrology and Water Resources, Technical Co-operation, and the Regional Offices.

Part of the satellite services under the space-based portion of the GOS is the forwarding of environmental products, and recent and anticipated advances promise increased efficiency within this function. For example, the METEOSAT Data Dissemination System (MDD) provides not only for the forwarding of products and data from the central processing centre to a remote readout station, but also for the return of vital environmental products.

WMO Technical Conference on the Economic and Social Benefits of Meteorological and Hydrological Services (Geneva, 1990)
The WMO Technical Conference on the Economic and Social Benefits of Meteorological and Hydrological Services was held in Geneva from 26 to 30 March 1990. It was attended by meteorologists, hydrologists, economists and engineers from national meteorological and hydrological

services, universities, private meteorological consulting firms and other interested institutions from all over the world. Papers presented covered (a) methodologies for assessing the economic and social benefits of meteorological and hydrological services, (b) user requirements for specific weather and climate services and related economic studies, (c) user requirements for hydrological services and related economic studies, and (d) the rôle and status of national meteorological and hydrological services in economic and social development.

The main conclusions were as follows:

* A concentrated combined further effort is needed to develop, evaluate, document and internationally exchange methodologies suitable for the assessment of economic and social benefits of meteorological and hydrological services.
* User requirements for such services are often highly specific and should be identified, defined and supplied through close co-operation between meteorologists/hydrologists, relevant users and intermediaries.
* Efforts are required to inform potential users of the benefits they can realize through making use of available improved services. Marketing of these services is increasingly important.
* Hydrological services need to be extended (where it is not yet done) to water-quality measurement and monitoring.
* The principle of the free exchange of basic meteorological information between national meteorological services should be preserved as the necessary basis for the provision of weather forecasts, warnings and other services all over the world.

WOCE see **World Ocean Circulation Experiment**

World Bank see **International Bank for Reconstruction and Development**

World Climate Applications and Services Programme (WCASP) see below

World Climate Applications Programme (WCAP)

The Climate Applications Programme of the World Climate Programme assists both developing and developed countries to collect, analyze and apply climate information to economic sectors such as agriculture, water resource use, energy generation and use, health, etc. There is close co-operation with the FAO in relation to the application of information on agriculture, forestry and fisheries, particularly with regard to agrometeorology and agroecology, and with FAO and a number of other agencies such as UNDP, UNDRO and UNEP in relation to meteorological and climatological aspects of locust control and desertification. There is also close co-operation with UNESCO and the International Association of Hydrological Sciences on water resources, while in the health sector,

WHO, Habitat and UNEP have all co-operated in studies to improve the habitability of urban complexes in the light of climate variations and climate change.

At the Eleventh Congress of WMO in May 1991 it was decided that the World Climate Applications Programme (WCAP) be changed to the World Climate Applications and Services Programme (WCASP). The WMO Congress considered that the new WCASP would reflect the increased emphasis on climatological services activities.

World Climate Conference (1979)

The (First) World Climate Conference, organized and convened by WMO, with the co-operation of other international organizations and member countries of WMO, took place in Geneva from 12 to 23 February 1979.

During the first week of the Conference, 26 overview papers on a variety of climate-related topics were presented and discussed by the world's leading specialists. The first week attracted over 350 experts from more than 50 countries. During the second week more than 100 experts from all parts of the world remained to engage in detailed discussions which resulted in a decision to issue a World Climate Conference Declaration in the form of an appeal to nations:

* to take full advantage of man's knowledge of climate;
* to take steps to improve that knowledge significantly;
* to foresee and to prevent potential anthropogenic changes in climate that might be adverse to the wellbeing of humanity.

The Declaration also contained a preamble, a discussion of the problem of climate and the future, and finally an appeal to all nations to support the World Climate Programme to be developed by WMO.

In his keynote address, entitled "Climate at the Millennium", Dr R. M. White drew attention to an important new concept arising from the Conference documentation. He pointed out that climate should be regarded as a resource even though it did not conform to the accepted definition of a resource. Furthermore, while access to climatic resources was restricted by national boundaries and property rights, climate also had some of the characteristics of a common property resource in that it could be modified by remote action. While the consequences of a global warming was essentially a matter of speculation at the time (1979), it was clear that such a change would have very different kinds of impacts in various regions of the world and that there would be winners and losers.

Taking an overall view of the Conference, there can be no question that it provided up to that time by far the most comprehensive assessment on the status of our knowledge of climate and its relationship to various aspects of the natural world and human society. The Conference was of the firm opinion that there was already a great deal of valuable information available about the natural variability of climate. It was urged that this knowledge could be put to immediate and continuing use in order to advance the economic progress of all nations and, especially, to establish climatic services in the developing countries.

There was considerable discussion of a proposal that the Conference should communicate to all governments and to the United Nations a strong recommendation for the convening of a global ministerial conference on matters related to climate. It was at length agreed, however, that such action should be deferred until research could reduce the existing uncertainties as to the future course of the global climate and provide more specific guidance on the social and economic impacts of anticipated changes in climate and its variability. Nevertheless, in view of the importance of exploiting to the full all available knowledge of climate variability, it was considered appropriate to suggest that regional conferences at ministerial level would serve many a valuable purpose. It was agreed that they could promote vigorous national and international action to stimulate the application of climatological data to various sectors of national economies, notably in developing areas.

World Climate Data and Monitoring Programme (WCDMP) see World Climate Data Programme (WCDP)

World Climate Data Programme (WCDP)

Although all parts of the WCDP are closely linked, the data component has particular importance because its objective is to ensure the timely availability of reliable climate data which are both accessible and exchangeable, in order to support climate applications, impact studies and research, and to assist developing countries in climate assessment for economic development.

WMO has taken the lead in creating services and projects within the WCDP. These include CLImate COMputer (CLICOM), Climate System Monitoring (CSM), and the Climate Change Detection Project (CCDP). Other agencies participating in the development of databases and services are UNEP through its Global Resource Information Database (GRID), ICSU through the World Data Centre Panel, and the Federation of Astronomical and Geophysical Services, and IOC through its international oceanographic data and information exchange system.

At the Eleventh Congress of WMO in May 1991 it was decided that World Climate Data Programme (WCDP) be changed to the World Climate Data and Monitoring Programme (WCDMP). The WMO Congress considered that the WCDMP would reflect the increased emphasis to climate monitoring activities.

World Climate Impact Assessment and Response Strategies Programme (WCIRP) see World Climate Impact Studies Programme (WCIP)

World Climate Impact Studies Programme (WCIP)

The lead agency for the Climate Impact Studies Programme is UNEP. The main objective of the WCIP is to improve the methodology of undertaking assessments of climate impacts. The various subprogrammes are co-ordinated by a Scientific Advisory Committee (SAC). UNEP in co-operation with the ICSU Scientific Committee on the Problems of the Environment (SCOPE) has prepared two publications on climate impact assessment and on atmospheric greenhouse gases and changing climate. With the assistance of the International Institute for Applied Systems Analysis (IIASA), UNEP has also prepared a study on policy-orientated assessment of the impact of climate variations and of the impacts of climate variations on agriculture.

At the Eleventh WMO Congress of WMO in May 1991 it was decided that the World Climate Impact Studies Programme (WCIP) be changed to the World Climate Impact Assessment and Response Strategies Programme (WCIRP). The WMO Congress considered that the WCIRP would embrace studies within both the "old" WCIP, as well as the current and future scientific and technical work on the identification of possible options for the mitigation and/or adaptation strategies to climate change and variability.

World Climate Programme (WCP)

The World Climate Programme is under the control of WMO, in association with UNEP, the Intergovernmental Oceanic Commission (IOC) of UNESCO, and ICSU. Initiated in 1979, as a result of the First World Climate Conference, it provides an institutional framework for research, applications, and data collection, specifically to improve the understanding of climate, and to assess its likely impacts. It has four specific components. From its initiation in 1989 until the meeting of the WMO Congress in May 1991, the components were as follows:

* World Climate Data Programme (WCDP) which assists countries in setting up climate data systems and acquiring processing capability in a way that could help economic policy-making.
* World Climate Applications Programme (WCAP) provides for the transfer of technology in the use of climate information to plan activities in agriculture, energy, transportation and human settlements, etc.
* World Climate Research Programme (WCRP) organized jointly by WMO and ICSU, with the support of IOC, is designed to understand the nature of the Earth's climate system in order to (a) increase the predictive capacity of weather/climate forecasts on a seasonal or monthly basis, and (b) assess the impact of human-induced influences on global climate change. It includes major projects such as Tropical Oceans and Global Atmosphere (TOGA), a study of the El Niño phenomenon, and the World Ocean Circulation Experiment (WOCE).
* World Climate Impact Studies Programme (WCIP) with UNEP as the lead agency has undertaken projects and set up expert groups to study the socio-economic impact of climate fluctuations and change. At the eleventh congress of WMO in May 1991 it was decided to change

the names and in some cases the terms of reference of the four components of the WCDP. The World Climate Data Programme (WCDP) would be changed to the World Climate Data and Monitoring Programme (WCDMP), the World Climate Applications Programme (WCAP) to the World Climate Applications and Services Programme (WCASP), and the World Climate Impact Studies Programme (WCIP) to the World Climate Impact Assessment and Response Strategies Programme (WCIRP). The fourth programme – the World Climate Research Programme (WCRP) – remained at it was. The WMO Congress considered that with respect to the WCDMP and the WCASP the new names would reflect the increased emphasis on climate monitoring and climatological services activities, and that the WCIRP would embrace studies within the "old" WCIP, as well as the current and future scientific and technical work on the identification of possible options for the mitigation and/or adaptation strategies to climate change and variability.

The World Climate Programme (WCP) promotes the use of climate information to assist economic and social planning and development, the improvement of the understanding of climate processes through internationally co-ordinated research, and the monitoring of climate variations or changes in order to be able to warn governments of climate impacts that may significantly affect human welfare and activities.

Since its inception, the World Climate Programme has aimed to achieve tangible results which could be used by nations in their operational activities and for national planning purposes. Projects within the Data, Application, and Impact Studies components have brought many useful results, such as expansion of the climate data-management computer systems, improving climate system monitoring, development of improved methods for applying climate information in various sectors, and assessing the impacts of climate change and sea-level rise. Within the World Climate Research Programme substantial scientific progress has been achieved, in particular, by refinements of global atmospheric circulation models, by studies towards improving an operational prediction capability for short-term climate variations, such as El Niño events, and by development of major experiments such as TOGA, WOCE and GEWEX.

World Climate Research Programme (WCRP)

The World Climate Research Programme is based on the successful co-operative experience of WMO and ICSU in organizing the Global Atmospheric Research Programme (GARP). The WMO–ICSU Joint Scientific Committee (JSC) has overall scientific responsibility for the Programme, with large inputs from both governmental services and the non- governmental academic communities. The purpose of the WCRP is to improve our knowledge of climate, climate variations and the mechanisms which bring about climate change so as to be able to determine to what extent climate can be predicted and the extent of man's influence on climate.

Within the framework of the WCRP, studies and experiments have been developed or are projected. These include (a) the development of improved numerical models of the atmosphere, capable of simulating the climate system, for carrying out climate predictions on a wide range of time and space scales; (b) a study of the tropical ocean and global atmosphere linkage (TOGA) in order to describe, model and predict the evolution of the coupled tropical oceans and global atmosphere system; (c) determination of the sensitivity of climate to various influences such as increasing concentrations of carbon dioxide and other greenhouse gases; and (d) the global energy and water cycle experiment (GEWEX) to observe, understand and model the processes of the global water cycle and energy budget.

World Food Conference (1974)
The World Food Conference was held in Rome in November 1974 and, although arranged independently, came to be regarded as a sequel – one of several – to the 1972 Conference on the Human Environment. Major factors of concern were the effects of variability in weather and climate in causing fluctuations in world food production. It was also recognized that efforts to increase food production throughout the world must take into account the weather and climate characteristics of each region.

The proceedings of the conference were conducted in two parts: the short-term problem of dealing with current food shortages; and the long-term requirements to increase food production and to institute a world food security system that would keep track of the worldwide situation and, whenever necessary, give early warning of any shortages likely to occur.

world forest conservation protocol or convention
A protocol or convention which may be developed in association with a climate convention which would cover boreal, temperate, subtropical and tropical forests.

World Health Organization (WHO)
The World Health Organization (WHO) was founded in 1946 by the International Health Conference that the Economic Social Council had convened in New York. The organization began operations in 1948 after 26 member states of the United Nations had ratified its constitution. The task of the WHO is to raise the health standards of all peoples to as high a level as possible. For this purpose, WHO maintains facilities all over the world for promotion of health. It co-operates with member states in the health sector and co-ordinates biological and medical research activities.

World Meteorological Organization (WMO)
The World Meteorological Organization (WMO), is a specialized agency of the United Nations. It currently has 160 Members (states and territories). It came into being in 1950 and its purposes are (a) to facilitate

worldwide co-operation in the establishment of networks of stations for the making of meteorological observations as well as hydrological and other geophysical observations related to meteorology, and to promote the establishment and maintenance of centres charged with the provision of meteorological and related services; (b) to promote the establishment and maintenance of systems for the rapid exchange of meteorological and related information; (c) to promote standardization of meteorological and related observations and to ensure the uniform publication of observations and statistics; (d) to further the application of meteorology to aviation, shipping, water problems, agriculture and other human activities; (e) to promote activities in operational hydrology and to further close co-operation between meteorological and hydrological services; and (f) to encourage research and training in meteorology and, as appropriate, in related fields and to assist in co-ordinating the international aspects of such research and training.

The Organization consists of the following:

* *The World Meteorological Congress (Congress)*. The Congress is the supreme body of WMO. It brings together the delegates of all Members once every four years (Congress last met in May 1991) to determine general policies for the fulfilment of the purposes of the Organization, to approve the WMO Long-term Plan, to authorize maximum expenditures for the following four-year financial period, to adopt technical regulations relating to international meteorological and operational hydrological practice, to elect the president and vice-presidents of the Organization and members of the executive council, and to appoint the secretary-general.

* *The Executive Council (EC)*. The Executive Council of WMO is composed of 36 elected members who are directors/general managers of national meteorological or hydrometeorological services. It meets once a year. Its main functions are to conduct the activities of the Organization, to implement decisions taken by Members in congress, and to study and make recommendations on any matter affecting international meteorology and related activities of the Organization.

* *The six Regional Associations* (I Africa, II Asia, III South America, IV North and Central America, V South-West Pacific and VI Europe), are composed of Members of WMO. They co-ordinate meteorological and related activities within their respective regions and examine from the regional point of view all questions referred to them.

* *The eight Technical Commissions*, consist of experts designated by Members. They are responsible for studying any subject within the purposes of the Organization. Technical commissions have been established for basic systems, instruments and methods of observations, atmospheric sciences, aeronautical meteorology, agricultural meteorology, marine meteorology, hydrology and climatology.

* *The Secretariat*, located in Geneva, Switzerland. It is composed of a Secretary-General (since 1984, the Secretary-General has been

Professor G. O. P. Obasi) and such technical and clerical staff as are required for the work of the Organization. It serves as the administrative, documentation and information centre of the Organization, makes technical studies as directed, supports all the bodies of the Organization, prepares, edits and arranges for the publication and distribution of the approved publications of the Organization, and carries out duties specified in the convention and other basic documents, and such other work as Congress, the Executive Council and the President decides. The Secretariat works in close collaboration with the United Nations and its specialized agencies.

World Meteorological Organization: early history

Several events led to the transformation of the International Meteorological Organization (IMO) into the intergovernmental World Meteorological Organization (WMO).

The first was in London in February 1946, when the Extraordinary Conference of Directors (of National Meteorological Services) took the first step "to bring IMO back into operation, to ensure its co-operation with other international organizations, and to resume the study of constitutional and other questions the settlement of which had been prevented by the war". The major task was to initiate action on preparing a new draft of the International Meteorological Convention.

The next key event was the Conference of Directors of National Meteorological Services in Washington DC in September–October 1947, at which the Convention of WMO was agreed to. The Convention came into force on 23 March 1950, that being the thirtieth day after the date of the deposit of the thirtieth instrument of ratification or accession. This date is celebrated annually as World Meteorological Day.

The First Congress of the World Meteorological Organization was convened in Paris in March 1951. It established the Executive Committee (now known as the Executive Council), technical commissions, regional associations and elected officers, and set forth the technical programme of the Organization to fulfil the purposes defined in Article 2 of the WMO Convention.

World Meteorological Organization: planning system

WMO Programmes are planned and managed within the framework of an integrated planning system that operates on three different over-lapping timescales.

* long-term plans (for ten years: 1984–93, 1988–97, 1992–2001) revised every four years which involve the identification of the main objectives which the Organization seeks to achieve, and the definition of overall policies, strategies and priorities for achieving them;
* medium-term plans (for four years: 1984–87, 1988–91, 1992–95, the beginning being coincident with that of each revised long-term plan), corresponding to the financial period of the Organization, which

provide the detailed budget for the first four years of the ten-year plan;
* short-term plans which are in very specific terms and closely co-ordinated with the biennial budget.

The formal introduction of long-term planning into WMO was effected through Resolution 34 of the Ninth World Meteorological Congress (1983) which adopted the First WMO Long-term Plan for the period 1984–93 and requested the Executive Council to establish appropriate machinery for the preparation of future long-term plans.

The Members of WMO participate in the long-term planning process through their involvement in sessions of the technical commissions and regional associations and through correspondence. The long- and medium-term plans are approved by all Members of WMO assembled at the four-yearly sessions of the World Meteorological Congress, while the short-term plans are approved biennially by the Executive Council. In all cases the period of effect of the plan commences approximately six months after its approval, at the beginning of the following calendar year.

World Ocean Circulation Experiment (WOCE)

The World Ocean Circulation Experiment (WOCE) is a worldwide oceanographic programme, organized as a component of the WCRP, to determine the oceanic circulation at all depths and in the global domain, within the period (1990–97).

The primary goal of WOCE is to develop global ocean models useful in the prediction of climate change and to develop the datasets necessary to test those models. Over the five years of the programme, there will be an intensification of the effort to determine air–sea fluxes globally by combining marine meteorological and satellite data, an upper-ocean measurement programme to determine the annual and interannual oceanic response to atmospheric forcing, and a programme of high-quality hydrographic observations. WOCE will also make intensive use of historical oceanographic data to assess the longer-term variability of ocean circulation.

In addition to observing activities at sea and in space, WOCE is planning a major oceanic modelling and data analysis activity, requiring a substantial increase in computer capabilities. Larger and faster computers are necessary to develop and run global eddy-resolving ocean models with sufficient vertical resolution, to simulate adequately the process of convection in the upper layers, the flow of deep water over sills and through passages, and the eddy exchanges of momentum, heat and salt in the strong boundary currents of the principal ocean gyres.

World Weather Watch (WWW)

The World Weather Watch (WWW) is a system created by WMO for collecting, analyzing and distributing throughout the world, weather and other environmental information. The WWW is an outstanding achievement in international co-operation: in few other fields of human endeavour has there ever been such a truly worldwide system, applying up-to-date scientific knowledge and technological developments, and to which virtually every country in the world contributes, every day of every year, for the common good.

WWW has three main components:

* the Global Observing System (GOS), comprising facilities on land, at sea, in the air, and in outer space for the observation and measurement of meteorological elements;
* the Global Telecommunication System (GTS) for the rapid exchange of observational information as well as analyses and forecasts produced by the third component;
* the Global Data-Processing System (GDPS), a network of computerized data-processing centres around the world.

These components are backed up by a variety of supporting activities, including the standardization of observing methods and techniques, the development of common telecommunication procedures and the presentation of both observational data and processed information in a manner understood by all, regardless of language.

The origins of WWW lie in the 1961 UN General Assembly Resolution 1721/XVI on the Peaceful Uses of Outer Space, which owed much to the address made by President J. F. Kennedy to the session of the General Assembly in September 1961 when he said :

"Scientists have studied the atmosphere for many decades but its problems continue to defy us. . . . With modern computers, rockets and satellites, the time is ripe to harness a variety of disciplines for a concerted attack . . . the atmosphere sciences require worldwide observation and hence, international co-operation . . . we shall propose further co-operative efforts between all nations in weather prediction . . . and . . . a global system of satellites linking the whole world."

Thus encouraged, the General Assembly passed a resolution calling on WMO to study measures that would: advance the state of atmospheric science and technology so as to provide greater knowledge of basic physical forces affecting climate and the possibility of large-scale weather modification; and develop existing weather-forecasting capabilities and help WMO Member countries make effective use of such capabilities through Regional Meteorological Centres. The programme was launched in 1963 by WMO's Fourth Congress.

In brief, WWW is the WMO's system which combines data-processing centres, observing systems and telecommunication facilities – operated by Members – to make available meteorological and related geophysical information needed in order to provide efficient meteorological and hydrological services within the countries. It also includes a Tropical Cyclone Programme, in which more than 50 countries are involved, and an Instruments and Methods of Observation Programme to promote standardization and development of meteorological and related observations.

The evolution of the WWW in the next decade will be guided by two main considerations.

* *Needs:* requirements of individual members for improved meteorological services as well as those of international research and applications programmes for meteorological and other environmental information;
* *Opportunities:* scientific achievements and technological advances as far as their operational inclusion in the WWW improves the quality of services provided.

Worldwide and long-range socio-economic development outlooks point to areas likely to affect user requirements for meteorological information. Three are of particular importance:

* pollution of the environment on a global scale and other possible changes of an anthropogenic origin to climate and the environment in general;
* continued and increasing problems of food, water and energy production and supply in many parts of the world;
* increased exploration and exploitation of resources in marine areas (oil, gas, minerals, fisheries, etc.).

The most widely recognized activity is likely to remain the supply of general and specialized weather forecasts for various user sectors, e.g. aviation, agriculture and water resources. There will, however, be changes with growing requirements for warnings of hazardous meteorological phenomena and very short-range forecasts with increasing specification of long-range weather forecasts or outlooks beyond ten days and up to a season. Not only will it be necessary to improve the quality of the present range of observational data but a wider range of atmospheric measurements will be required. It is very likely also that much greater use will be made of the WWW infrastructure for exchanging information and ensuring warnings of a meteorological nature.

The WWW provides both the common infrastructure and the database to support a broad range of WMO programmes and relevant efforts of international organizations. These include the World Climate Programme (WCP), the World Area Forecast System (WAFS), the Integrated Global Ocean Services System (IGOSS), and the IAEA Conventions regarding the release of hazardous materials in the atmosphere. A continued dialogue with those responsible for these programmes is therefore essential to ensure that changes in requirements are properly incorporated into the WWW Plan.

World Weather Watch Data Management (WWWDM)

The WWW Data Management (WWWDM) is the component within the WWW system which provides those support functions needed for the orderly overall management of meteorological data and products of the WWW system, the most economical use of the resources of the WWW system components, and for monitoring data and product availability and quality. The underlying principle in the WWWDM design is the need for the integration of the GOS, GTS and GDPS subsystems, facilities, services and functions into an efficient system.

WWW see **World Weather Watch**

WWWDM see **World Weather Watch Data Management**

X

xerophyte
A plant that can grow in dry places.

Z

zonal
Along a line of latitude; that is west–east or east–west (see also **meridional flow**)

zonal index
A circulation index relating to the west to east component of the atmospheric circulation.

zooplankton
The portion of the plankton community comprised of tiny aquatic animals eaten by fish.